Mind Maps in Clinical Chemistry for Core Competencies in Laboratory Diagnostics

Authored by

Simmi Kharb
Department of Biochemistry
Pt. B.D. Sharma Post Graduate Institute of Medical Sciences (PGIMS)
Rohtak 124001, Haryana, India

Mind Maps In Clinical Chemistry for Core Competencies in Laboratory Diagnostics

Author: Simmi Kharb

ISBN (Online): 979-8-89881-651-3

ISBN (Print): 979-8-89881-652-0

ISBN (Paperback): 979-8-89881-653-7

Published by Bentham Science Publishers Pte. Ltd. Singapore,

in collaboration with Eureka Conferences, USA.

First published in 2026.

Bentham Science Publishers Pte. Ltd.
No. 9 Raffles Place
Office No. 26-01
Singapore 048619
Singapore
Email: subscriptions@benthamscience.net

BENTHAM SCIENCE

CONTENTS

PREFACE

The book *Mind Maps in Clinical Chemistry for Core Competencies in Laboratory Diagnostics* is for students studying clinical biochemistry as part of their medical, nursing, or biomedical science courses and for those preparing for postgraduate entrance and board examinations. The book has unique learning modules designed on the outline of clinical chemistry and laboratory evaluation of disease with a focus on clinical cases. It will serve as a companion for the core competencies for the general clinical chemistry exam conducted by various Clinical Chemistry Boards.

The outline of this book has the headings of: *Key concepts; biochemical basis, relevant lab tests, interpretation and applied aspects; relevant laboratory tests for detection, diagnosis, management, and monitoring; relevant laboratory tests, analytics methods available and applied aspects*.

This book is a learning tool for imparting mastery and knowledge of clinical chemistry interpretation and problem-solving for preparing for exams and board certifications.

Simmi Kharb
Department of Biochemistry
Pt. B.D. Sharma Post Graduate Institute of Medical Sciences (PGIMS)
Rohtak 124001, Haryana, India

CHAPTER 1

Disorders of Acid-Base and Pulmonary Function

Abstract: The chapter discusses the critical clinical conditions related to acid-base balance and pulmonary function. It emphasizes the importance of identifying and managing acid-base disorders through ABG interpretation and serum electrolyte analysis. Additionally, it highlights the essential role of Pulmonary Function Tests (PFTs) in diagnosing and managing respiratory conditions such as COPD and asthma.

LEARNING OBJECTIVES

After reading this chapter, the reader should be able to:

1. Describe the principles and control of acid-base balance and pulmonary function
2. Discuss the causes, clinical signs, and symptoms of the disorders relating to acid-base balance and pulmonary function
3. Describe the laboratory investigations that are important in their detection, diagnosis, and management
4. Describe the specific laboratory investigations important to the study of acid-base balance and pulmonary function
5. Understand the principle of their measurement
6. Selection, interpretation, and clinical correlation of laboratory investigations

KEY CONCEPTS

Key Concepts: Acid-base Regulation and Pulmonary Function:

Principles and control of acid-base balance and pulmonary function

Acid-base balance and pulmonary functions [1,2]

Acid-base balance

Regulated tightly.

Arterial blood pH: 7.38-7.42, kept constant by buffer solutions.

Simmi Kharb

Acid-base homeostasis

The body is highly sensitive to changes in its pH levels.

Body pH is maintained.

To maintain this, strong mechanisms are in place.

If pH is outside the acceptable range, there is:

Protein denaturation and digestion

Loss of the ability of an enzyme to function

Compromised ability of the body to maintain homeostasis, making survival incompatible with life

Chances of death.

Buffer

A mixture of a weak acid and its conjugate base, *or*

A mixture of a weak base and its conjugate acid

When a small amount of strong acid or base is added to it, very little change occurs in the pH.

Extracellular buffers include: bicarbonate, ammonia.

Intracellular buffers include: proteins, phosphates.

Buffer systems and various organs maintain extracellular pH

These systems maintain a pH value of 7.36-7.44.

The respiratory system modulates pCO_2.

Kidneys modulate the concentration of bicarbonate.

Buffers

Bicarbonate

This helps regulate and maintain optimal pH levels.

This regulates the concentration of carbon dioxide.

Le Chatelier's principle

When a chemical system at equilibrium experiences a *change in concentration, temperature, or total pressure*, its equilibrium shifts to minimize the change.

In the context of acid-base imbalances:

Acid-base imbalances are *overcome* by short-term compensation *via*:

A change in the ventilation rate alters the blood carbon dioxide to alter pH.

Renal compensation

Slowest to compensate.

The kidney *controls pH* by *excreting excess acid or base.*

Acidosis

Compensation by renal tubules:

Bicarbonate is absorbed by renal tubular cells.

Compensation by collecting duct:

Hydrogen secretion.

Bicarbonate generation.

Ammoniagenesis: increase in NH_3 buffer.

Alkalosis

Bicarbonate excretion.

Compensation by renal tubules:

Decreased hydrogen ion secretion.

Decreasing glutamine metabolism: ammonium excretion decreased.

Regulation of pH by lungs

Via changes in the rate of respiration to:

Expel CO_2.

Raise pH and restore normal pH.

Mechanism:

Blood carries hydrogen ions, oxygen, and carbon dioxide.

CO_2

60% of CO_2 is carried as *dissolved bicarbonate.*

Only a small amount of CO_2 is transported as *carbaminohemoglobin*: transported to the lungs and removed.

Increased rate of ventilation:

pH restored to normal.

Compensation for acidemia:

Expulsion of CO_2.

Compensation for alkalemia:

Retention of CO_2.

Examples

i. Smoking affects blood pH

Cigarette smoking increases pH.

Chronic conditions (COPD) can occur due to heavy smoking:

In COPD, maintenance of a balanced pH becomes difficult.

Smoking increases blood CO levels, and CO displaces oxygen in RBCs by forming COHb%.

Carboxyhemoglobin shifts the hemoglobin dissociation curve to the left:

Reduced the capacity of hemoglobin to release oxygen to tissues.

Higher hemoglobin levels in smokers compensate for reduced oxygen delivery compared with non-smokers.

ii. Urine testing

Acid-base imbalances can be detected by urine testing.

iii. Uncontrolled diabetes

Highly acidic urine is passed.

Kidneys can be over-stressed and malfunction.

This could result in a coma or death.

CAUSES, CLINICAL SIGNS, AND SYMPTOMS OF THE DISORDERS RELATING TO ACID-BASE BALANCE AND PULMONARY FUNCTION

Imbalance of acid-base

In an acid-base imbalance, blood pH shifts from its normal range (7.35 to 7.45).

Acidemia: Excess of acid in the blood.

Alkalemia: Excess of the base in the blood.

Types

Classification:

According to the etiology:

Respiratory

Metabolic

According to the change in pH:

Acidosis

alkalosis

The following conditions can occur solely or in mixed combination:

Metabolic acidosis

Respiratory acidosis

Metabolic alkalosis

Respiratory alkalosis

COMPENSATORY MECHANISM FOR MAINTAINING BALANCED BLOOD PH

Acid–base imbalances *overcome the buffer system*

At the level of the lungs

Compensation occurs through changes in the rate of ventilation, which alter blood carbon dioxide concentration and shift the above reaction to adjust pH.

Le Chatelier's principle governs the basic reaction:

$H_2O + CO_2 \leftrightharpoons H_2CO_3 \leftrightharpoons \mathbf{H^+} + CO_3^-$

Hydrogen ions are free, and pH will rise back to normal.

For alkalemia, the opposite occurs.

At the level of the kidneys

Features:

Kidneys are slower to compensate than the lungs.

Renal physiology has several powerful mechanisms to control pH by excretion of excess acid or base.

Renal excretion is a major homeostatic control point for maintaining a stable pH balance.

Mechanism

Homeostasis occurs *via* the excretion of waste products.

Excrete H^+ ions into urine.

Reabsorb bicarbonate from urine.

During acidosis

Increased resorption of bicarbonate from the tubular fluid.

Hydrogen secretion in the collecting ducts generates more bicarbonate.

Increased formation of NH_3 buffer.

During alkalosis

Causes increased excretion of bicarbonate by the kidney.

Reduced secretion of hydrogen ions occurs.

Increased excretion of ammonium.

BUFFER SYSTEMS IN THE HUMAN BODY

Bicarbonate buffer (HCO_3–/CO_2), haemoglobin buffer (in erythrocytes), phosphate buffer, proteins, and ammonium buffer.

BICARBONATE (HCO_3^-)

Most important buffer.

Fastest

Under ideal conditions in the human body:

At physiological pH 7.4, pKa 6.1, HCO_3 is tenfold more than CO_2

Bicarbonate buffers effectively at pH 7.4, given its pKa of 6.1.

Bicarbonate to carbonic acid is present in a ratio 20:1.

Higher concentration of bicarbonate neutralizes a significant amount of acidity.

Dissolved CO_2 in blood is neutralized by the *bicarbonate-carbon dioxide buffer system* [a weak acid, carbonic acid (H_2CO_3)].

H_2CO_3 dissociates into a hydrogen ion and a bicarbonate ion.

This buffer system allows the body to maintain physiologic pH.

Carbonic anhydrase catalyzes the conversion of CO_2 and water to H^+

and bicarbonate:

$CO_2 + H_2O \rightarrow H^+ + HCO3^-$

Maintains acid-base balance in the bloodstream

Located in red blood cells, renal tubules, gastric mucosa, and pancreatic cells.

Present in high concentrations in erythrocytes:

> 70% of produced HCO_3^- leave erythrocyte using special *HCO_3^-/Cl^- antiport*

Bicarbonate is exchanged for Cl^-: *Hamburger's effect, chloride shift.*

CO_2:

Bound to hemoglobin → carbamino compound

When CO_2 and H^+ concentrations are high:

Affinity of hemoglobin for O_2 is decreased.

When CO_2 concentrations are low:

Affinity of hemoglobin for O_2 is increased: ***Bohr effect.***

If O_2 concentrations are high:

Increased unloading of CO_2 from the tissues: ***Haldane effect***

As levels of CO_2 in the blood begin to rise, the body can respond through hyperventilation.

Sources of CO_2 in the body

Metabolic:

Cellular respiration:

Metabolic alkalosis.

Cushing's syndrome.

Kidney failure.

- Stomach and pancreas produce sodium bicarbonate.
- * Proximal and collecting tubules reclaim bicarbonate by hydrogen excretion.

Mechanical:

Chronic Obstructive Pulmonary Disease (COPD).

Certain neurological and muscle diseases.

Metabolic disruptions.

Damage to areas of the brain that regulate breathing.

Respiratory:

Lung or breathing disorders.

Hyperventilation.

Homeostatic mechanism:

Does not have a transporter.

Its reabsorption involves a series of reactions in the tubular lumen and the tubular epithelium.

At the renal level

In response to acidosis:

Tubular cells reabsorb more bicarbonate from the tubular fluid.

Collecting duct cells:

Secrete more hydrogen.

Generate more bicarbonate.

Ammoniagenesis→ Increased formation of NH_3 buffer.

In response to alkalosis:

Kidneys may excrete more bicarbonate:

By decreasing hydrogen ion secretion from tubular epithelial cells.

Kidneys lower rates of glutamine metabolism and ammonium excretion.

At the level of the lungs

Acidemia:

Blood pH drops.

Compensated by increasing the rate of respiration to expel CO_2.

The above reaction shifts to the left:

Fewer hydrogen ions.

The pH returns to normal.

For alkalemia:

The mechanism is the opposite in alkalemia.

PH OF THE HUMAN BODY AVERAGES AT 7.40

The range of pH of the human body: 7.35- 7.45

If pH is below 7.35, acidemia occurs

If pH is above 7.45, alkalemia occurs.

pH at a 7.4 level is ideal for many biological processes:

One of the most important is the oxygenation of the blood.

Many intermediates of biochemical reactions in the body become ionized at a neutral pH:

This causes their utilization to be more difficult.

To maintain a pH level within the required narrow range, the human body contains compensatory mechanisms.

Counterbalance homeostasis in acid-base disorders

Four main types of acid-based disorders experienced by the human body:

Metabolic acidosis, metabolic alkalosis, respiratory acidosis, and respiratory alkalosis

If one of these conditions occurs:

A counterbalance is induced in the form of an opposite condition.

In case of metabolic acidemia:

The body attempts to induce a respiratory alkalosis to compensate.

NOTE: It is rare for the compensation to restore the pH to 7.4.

Phosphate buffer system

Important for the regulation of urine pH: the *most predominant urine buffer.*

Its urinary excretion increases with acidosis.

NOT an important blood buffer: its concentration is too low.

Found in the blood in two forms:

Sodium dihydrogen phosphate ($Na_2H_2PO_4^-$): weak acid.

Sodium mono hydrogen phosphate ($Na_2HPO_4^{2-}$): weak base.

Urine buffers

Ammonium buffer (NH_3/NH_4^+).

Phosphate buffer.

Titratable urine acidity:

20 mmol of phosphates is excreted per day.

Physiologic urine pH: 4.4- 8.0.

Ammonia excretion

30-50 mmol of NH_4^+ is excreted per day.

Significantly regulated: could be decreased or increased.

In acidosis:

Glutaminase is activated in the kidneys.

Glutaminase splits glutamine into glutamate and NH_3.

NH_3 is eliminated in urine.

In the liver, less urea and more glutaminase are produced.

Protein buffers

Plasma proteins and intracellular proteins constitute the most abundant and powerful buffer system.

Hemoglobin: most important among blood proteins.

35% of the buffering capacity of blood is provided by Hb.

Only 7% of the buffering capacity of the blood is provided by the remaining proteins.

Proteins assist in intracellular pH regulation.

Role of erythrocytes and haemoglobin in the acid-base balance

CO_2 diffuses to erythrocytes.

In RBC, CO_2:

Either binds to haemoglobin to form carbaminohemoglobin or

Reacts with water, catalyzed by carbonic anhydrase.

Organ systems involved in acid-base balance [3,4]

An optimal pH balance is essential for the proper functioning of every organ system in the human body.

Two main modulatory systems:

Renal

Pulmonary

Pulmonary system adjusts pH:

Via regulation of carbon dioxide levels

Compensating for Metabolic pH disturbances

It occurs in minutes to hours.

The renal system restores pH by:

Reabsorbing bicarbonate: pH becomes more alkaline

Excreting fixed acids

This process takes several days.

LABORATORY INVESTIGATIONS THAT ARE IMPORTANT IN THEIR DETECTION, DIAGNOSIS, AND MANAGEMENT

Related Testing

1. Arterial Blood Gas (ABG) sampling

Done to determine parameters: ***pH, HCO_3, pCO_2, pO_2, oxygen saturation*** by using *Ion-Selective Electrodes (ISE).*

Normal values of ABG:

pH = 7.35 -7.45

pCO_2 = 35- 45 mmHg

pO_2 = 75- 100 mmHg

HCO_3^- = 22- 26 mEq/L

O_2 Sat = > 95%

2. Other tests

Parameters for acid-base status:

Electrolytes: Cations: [Na^+], [K^+], [$Ca2^+$], [$Mg2^+$]; Anions: [Cl^-].

Renal function

Anions: [Cl^-], [lactate], albumin

Metabolites: [urea], [creatinine], [ketone bodies]

They assist in determining the mechanisms and contributing factors of acid-base imbalance disorders.

Anion gap

Within any body fluid compartment, anions and cations are equal to each other (*e.g.*, plasma) under normal conditions.

Major cation: Na^+

Major anions: $HCO3^-$, Cl^-

Measured Na^+ is greater than HCO_3- and Cl^-, and there are unmeasured anions that maintain electroneutrality. Unmeasured plasma anions include plasma proteins, citrate, phosphate, and sulfate.

Formula

Anion gap $= Na^+ - (Cl^- + HCO_3^-)$

Normal values = 8-16 mEq/L

Use

Very useful for narrowing down the differential diagnosis of metabolic acidosis:

Normal anion gap metabolic acidosis (*or hyperchloremic metabolic acidosis with a normal anion gap).*

Hyperalimentation

Diarrhea

Saline infusion

Addison's disease

Renal tubular acidosis

Acetazolamide

Spironolactone

Elevated anion gap metabolic acidosis

Salicylates

Methanol

Propylene glycol

Ethylene glycol

Iron tablets

Isoniazid

Uremia

Diabetic ketoacidosis

Lactic acidosis

Compensation for Acidosis and Alkalosis

In case of acid-base imbalance (Table **1**), there are two options in the body:

Either to *fix the original problem* or,

To provide a *temporary fix* till the permanent fix comes into action.

Hypoventilation raises pCO_2, decreases pH ($[H^+]$ increased), causing *respiratory acidosis*.

Compensation

Hyperventilation does not seem to be an option.

The strategy should be to change the bicarbonate concentration to get the pH back to normal, which leads to *metabolic alkalosis*.

Table 1. Disorders of acid-base balance.

Condition	PCO_2	H^+	$[HCO_3^-]$
Respiratory acidosis	↑	↑	↓
Respiratory alkalosis	↓	↓	↓
Metabolic acidosis	↓	↑	↓
Metabolic alkalosis	↑	↓	↑

Metabolic cause is **compensated by** respiratory; acidosis is **compensated by** alkalosis.

Control of respiration[3]

Respiration is controlled by areas of the brain that stimulate the contraction of the diaphragm and intercostal muscles. The goal of the system is to keep the pH of the bloodstream within normal neutral ranges, around 7.35.

Respiratory centres

Medullary inspiratory centre

Located in the medulla oblongata.

Stimulates contraction of inspiratory muscles (diaphragm and external intercostal muscles).

Pheumotaxic area

Located in pons.

Inhibits the inspiratory centre: limits the contraction of the inspiratory muscles and prevents the lungs from overinflating.

Apneustic area

Located in pons.

Stimulates the inspiratory centre and prolongs the contraction of the inspiratory muscles.

Chemoreceptors and respiratory centres

Influenced by stimuli received from two groups of sensory neurons:

Central chemoreceptors

Located in the medulla oblongata.

Monitor the chemistry of cerebrospinal fluid.

Entry of plasma CO_2 in CSF

Forms HCO_3^- and H^+.

Lowers the *pH* of fluid (becomes more acidic).

In response to lowered pH, central chemoreceptors *stimulate the respiratory centre* to increase the inspiratory rate.

Peripheral chemoreceptors

Located in aortic bodies in the wall of the aortic arch and in carotid bodies in the walls of the carotid arteries.

Monitor the chemistry of blood.

Increase in pH or pCO_2, or a decrease in pO_2

Stimulates these receptors, which in turn *activate the respiratory centre.*

These receptors signal the respiratory centre to *discontinue stimulation of the inspiratory muscles*, allowing expiration to begin.

Chemoreceptor regulation of breathing

Chemoreceptors detect blood CO_2 levels by monitoring blood $[H^+]$

Increase in $[CO_2]$→ decrease in blood pH → central chemoreceptors *stimulate the respiratory centre* to increase the inspiratory rate

Hyperventilation → alkalosis→ feedback response of *decreased ventilation* (to increase CO_2)

Hypoventilation →acidosis→feedback response of *increased ventilation* (to remove CO_2).

Any situation with *hypoxia* (too low O_2 levels) → feedback response →increases ventilation → increases oxygen intake.

Appropriate respiratory feedback response is generated in:

Vomiting → alkalosis

Diarrhea → acidosis.

Henderson-Hasselbach equation

Provides a relationship between the pH of acids (in aqueous solutions) and their pKa (acid dissociation constant).

The equation is fundamental in acid-base chemistry for diagnosis and managing conditions such as acidosis or alkalosis.

Equation

$pH = pKa + \log([HA]/[A^-])$

Where:

pH: hydrogen ion concentration (acidity).

pKa: negative logarithm of acid dissociation constant (Ka) of a weak acid.

$[A^-]$: concentration of the conjugate base.

[HA]: concentration of the undissociated weak acid.

For the bicarbonate buffer system:

pKa of carbonic acid (H_2CO_3): 6.1

The ratio of bicarbonate ions (HCO_3^-) to carbonic acid (H_2CO_3) maintains blood pH around 7.4.

Hemoglobin dissociation curves and calculated oxygen saturation

Oxygen Delivery to Tissues

Oxygen dissociation curve:

Depicts the relationship between the partial pressure of oxygen and the saturation of hemoglobin.

Provides information on the ability of hemoglobin to deliver oxygen to tissues.

Shifting of the O_2 dissociation curve to the left:

p50 decreased, causing pH to be in the alkalotic range.

Implies: increased affinity of hemoglobin for oxygen.

Shifting of the O_2 dissociation curve to the right:

Due to a decrease in pH.

Implies: decreased affinity of hemoglobin for oxygen.

<u>Osmolar gap</u>

Osmole

Amount of a substance that yields under ideal solution number of particles that would depress the freezing point of the solvent by 1.86 K.

Osmolality of a solution

Number of osmoles of solute per kilogram of solvent (mOsm/kg of solute).

Osmolarity of a solution

Number of osmoles of solute per litre of solution.

Calculated osmolarity

(2 x [Na^+]) + [glucose] + [urea])

Osmolar gap

Osmolality (measured) – Osmolarity (calculated)

normal = < 10

High osmolar gap

Presence of an abnormal, uncharged solute with low molecular weight present in significant amounts.

Serum osmolar gap > 10 mOsm/l:

Considered abnormal.

Indicates the presence of an osmotically active substance in the blood.

Causes:

Due to the presence of elevated levels of other osmotically active particles:

Methanol

Ethylene glycol

Polyethylene glycol

Propylene glycol (iv lorazepam, diazepam, and phenytoin)

Glycine

Maltose

Mannitol

Sorbitol.

Types of Osmometers

Osmometer is used to measure osmolality.

Use the colligative property of freezing point depression.

Use the colligative property of vapour pressure depression.

Elevated osmolar gap indicates an unknown solute, but does not identify it

Example:

Methanol ingestion: increased osmolar gap; the size of the osmolar gap is proportional to the amount of toxin.

Ethanol, as well as ethylene glycol or methanol ingestion:

Ethanol will increase the osmolar gap.

There are chances of missing the presence of more toxic substances.

Other common causes of elevated osmolar gap:

Alcoholic ketoacidosis, kidney failure, diabetic ketoacidosis, and shock.

Decreased water in the blood or increased solutes: Dehydration.

Shock

Diabetes

Increase blood glucose

Increased blood sodium

Increased nitrogen waste products in the blood

Stroke or head trauma that leads to decreased Antidiuretic Hormone (ADH)

Kidney damage and disease.

Urine Osmolar Gap:

Represents ammonium salts = Measured Urine Osmolality - (2 x Na + 2 x K + Urea/2.8)

Urine Osmolar Gap / 2 < 150 -> RTA 1 and 4

Urine Osmolar Gap / 2 > 400 -> Look to the gut

Systematic approach to investigating acid-base disturbances[4]

Sequence of interpretation of ABG report:

See: **pH** ↓

Try to **decide** whether:

Acidotic

Alkalotic

Within physiological range ↓

See: **$PaCO_2$ levels** to determine respiratory contribution:

High levels ⟶ Respiratory compensation work by decreasing pH when needed and increasing it when necessary

See: **HCO_3^- levels** to determine metabolic status/kidney effect: ↓

Elevated HCO_3^- raise pH

and *vice versa*.

See: If H is **acidotic**: ↓

Identify a value associated with a **lower pH:**

In respiratory acidosis: High CO_2.

In metabolic compensation: High HCO_3^-

In metabolic acidosis: Low HCO_3^-

See: In case pH is **alkalotic**: ↓

Identify the value responsible for this:

Respiratory alkalosis: Low CO_2

Metabolic alkalosis: High $HCO3^-$

See for a **compensatory mechanism***that will be inversely reflected* ↓ :

Look for values:

Respiratory alkalosis: Low HCO_3^- is the *metabolic response*

Metabolic alkalosis: high CO_2 is the *respiratory response*

See for **mixed disorders**: ↓

Look for values:

pH levels: in the physiological range

$PaCO_2$ and/or HCO_3^- levels: not within normal limits

DISEASES

Alpha-1 Antitrypsin (A1AT) deficiency

(Also read hepatobiliary disease)

Biochemical basis of:

Causes and clinical signs and symptoms:

Causes:

Defect:

Alpha-1 Antitrypsin Deficiency (AATD)

Mutation in the SERPINA1 gene on chromosome 14q32

Inheritance: autosomal codominant

Clinical signs and symptoms:

Shortness of breath (on exertion and later at rest)

Excessive cough with phlegm/sputum production

Wheezing

Decrease in exercise capacity and a persistent low energy state or tiredness

Chest pain that increases when breathing in

Yellow discoloration of skin

Weight loss

Jaundice

Diagnostic criteria:

Alpha-1 Antitrypsin Deficiency (AATD) is suspected in:

Chronic obstructive pulmonary disease

And/Or:

Liver disease at any age, including obstructive jaundice in infancy. In adults, anyone with unexplained liver abnormalities, *e.g.*, elevated liver enzymes.

C-ANCA positive vasculitis (*i.e.*, granulomatosis with polyangiitis)

Necrotizing panniculitis

Low serum concentration of Alpha-1 Antitrypsin (AAT).

Biochemical changes of disease or disorder:

Alpha-1-Antitrypsin (AAT) belongs to the serine proteinase inhibitor (serpin) family.

In α-1 Antitrypsin (AAT) deficiency, misfolded AAT accumulates in the endoplasmic reticulum of hepatic cells and other cells that produce AAT.

Following the loss of AAT, the lung protease-antiprotease balance gets disrupted. The proteases (specifically neutrophil elastase) destroy the lung matrix, alveolar structures, and cause Chronic Obstructive Pulmonary Disease (COPD), emphysema ensues, and it also causes liver damage in 15% of cases.

Relevant laboratory tests:

For detection, monitoring, and management:

Low serum concentration of Alpha-1 Antitrypsin (AAT)

SERPINA1 variants (Biallelic pathogenic variants)

Protease Inhibitor (PI) typing to detect dysfunctional AAT protein variant.

Assessment of individual risk for disease:

α1-Antitrypsin Deficiency (AATD)

Reduced levels of α1-Antitrypsin (AAT).

The most common deficiency alleles associated with reduced serum AAT levels include PI*Z and PI*S, and AATD is most strongly associated with the Z variant.

Low serum AAT levels are seen in homozygotes with severe deficiency alleles (*i.e.*, PI*ZZ)

They have an increased risk of developing Chronic Obstructive Pulmonary Disease (COPD)

Individuals with alleles (*e.g.*, Z allele, Mmalton allele, Siiyama allele) are at increased risk of developing liver disease.

Smoking is a key risk factor for the development of lung disease in patients with AATD, and disease progression and survival are both significantly worse in smokers than in non-smokers.

The presence/extent of emphysema and its association with allele variants should be assessed.

Interpretation of analytical results:

In terms of clinical signs and symptoms of the disease

Normal serum AAT levels:

20-53 μmol/L by the enzymatic method

100-220 mg/dL by nephelometry.

Serum AAT levels in lung disease are diagnostic if <57 mg/dL.

Analytical methods:

Available for the laboratory tests:

ELISA, nephelometry: Alpha-1 Antitrypsin (AAT)

Electrophoresis:

SERPINA1 alleles

Polyacrylamide gel Isoelectric Focusing (IEF) electrophoresis:

Dysfunctional AAT protein variant by Protease Inhibitor (PI) typing

Molecular genetic testing:

Gene-targeting:

Single-gene testing

Multigene panel for alleles: PI*Z, PI*S, PI*I, and PI*F

Comprehensive genomics:

Exome sequencing

Genome sequencing.

Sequence analysis: SERPINA1

Carbon monoxide poisoning

Biochemical basis:

Causes and clinical signs and symptoms:

Causes:

CO has 240 times higher affinity for Hb than O_2 and binds tightly to Hb

Thus, it forms Carboxyhemoglobin (COHb)

Outcome: COHb binding reduces oxygen transport and shifts the oxygen dissociation curve toward the left.

Clinical signs and symptoms:

The most common symptoms of CO poisoning are:

Headache, dizziness, weakness, nausea, vomiting, chest pain, and confusion

Biochemical changes of disease or disorder:

CO produces toxicity by binding to hemoglobin, thereby reducing oxygen-carrying capacity, and lactic acidosis (tissue hypoxia).

Relevant Laboratory Tests:

For detection, monitoring, and management:

CO-oximetry

Arterial Blood Gas (ABG), COHb

Serum lactate

Interpretation of analytical results

COHb >10%(smokers) or

COHb **>3% (non-smokers)** confirms CO poisoning

Lactic acidosis supports tissue hypoxia.

Analytical methods:

Available for laboratory tests

Co-oximetry (spectrophotometry-based detection of COHb)

Gas chromatography for CO levels

Blood gas analysis for acid-base disturbances.

Metabolic acidosis

Biochemical basis

The accumulation of acids or the loss of bicarbonate leads to a decrease in blood pH.

Common causes: diabetic ketoacidosis, renal failure, lactic acidosis

Clinical manifestation.

Rapid breathing (Kussmaul respiration), confusion, fatigue, nausea, and shock.

Biochemical changes in the disease or disorder.

Low bicarbonate, decreased pH, high anion gap (if due to organic acids).

Relevant laboratory tests:

Arterial Blood Gas (ABG)

Serum bicarbonate

Anion gap

Serum lactate

Ketones.

Interpretation of analytical results

Low pH and HCO_3^- confirm acidosis

Anion gap helps differentiate causes

Analytical methods:

Available for laboratory tests

Blood gas analysis

Enzymatic assays for lactate and ketones

Metabolic alkalosis

Biochemical basis:

Excess bicarbonate due to loss of acid (vomiting, diuretics) or bicarbonate retention (hyperaldosteronism)

Causes and clinical signs and symptoms:

Muscle cramps, tetany, confusion, arrhythmias, hypoventilation

Biochemical changes of disease or disorder:

Increased bicarbonate, increased pH, and hypokalemia.

Relevant laboratory tests:

ABG, serum bicarbonate, chloride levels, and urine electrolytes.

Interpretation of analytical results:

Elevated pH and HCO_3^- confirm alkalosis

Urinary chloride helps identify etiology

Analytical methods:

Ion-Selective Electrode (ISE) for electrolytes, blood gas analysis.

Pyloric stenosis

Biochemical basis:

Gastric outlet obstruction leads to persistent vomiting, causing metabolic alkalosis and dehydration.

Clinical Manifestations:

Projectile vomiting, failure to thrive, dehydration, palpable "olive-shaped" mass.

Biochemical Changes:

Hypochloremic metabolic alkalosis, hypokalemia, increased bicarbonate.

Relevant laboratory tests:

Serum electrolytes, blood gas, and ultrasound imaging.

Interpretation of analytical results:

High HCO_3^- and pH, low chloride and potassium confirm metabolic alkalosis.

Analytical methods:

Blood gas analysis and ultrasonography.

Renal tubular acidosis (also refer to disorders of the kidney and urinary tract)

Biochemical basis:

Impaired renal acid excretion or bicarbonate reabsorption leads to acidosis.

Types: Distal (Type 1), Proximal (Type 2), Hyperkalemic (Type 4).

Causes, clinical signs, and symptoms.

Biochemical changes of disease or disorder.

Clinical Manifestations:

Growth retardation, muscle weakness, nephrocalcinosis, and kidney stones.

Biochemical Changes:

Normal anion gap metabolic acidosis, hyperchloremia, and variable potassium levels.

Relevant laboratory tests:

ABG, urine pH, serum electrolytes, urinary anion gap, ammonium excretion test.

Interpretation of analytical results:

Urine pH >5.5 in Type 1 RTA, low bicarbonate in Type 2 RTA.

Analytical methods:

Blood gas analysis, urine electrolyte assays, ammonium chloride loading test.

Respiratory alkalosis:

Biochemical basis:

Impaired renal acid excretion or bicarbonate reabsorption leads to acidosis. Types: Distal (Type 1), Proximal (Type 2), Hyperkalemic (Type 4).

Clinical Manifestations:

Growth retardation, muscle weakness, nephrocalcinosis, and kidney stones

Biochemical Changes:

Normal anion gap metabolic acidosis, hyperchloremia, variable potassium levels.

Relevant laboratory tests:

ABG, urine pH, serum electrolytes, urinary anion gap, ammonium excretion test.

Interpretation of analytical results:

Urine pH >5.5 in Type 1 RTA, low bicarbonate in Type 2 RTA.

Analytical methods:

Blood gas analysis, urine electrolyte assays, and ammonium chloride loading test.

Respiratory acidosis

Biochemical basis:

CO_2 retention due to hypoventilation (*e.g.*, COPD, neuromuscular disorders).

Clinical Manifestations:

Shortness of breath, confusion, drowsiness, headache, and cyanosis.

Biochemical Changes:

Increased CO_2 (hypercapnia), low pH, and normal/high bicarbonate (compensation).

Relevant laboratory tests:

ABG, serum bicarbonate, and pulmonary function tests.

Interpretation of analytical results:

Low pH and increased pCO_2 confirm respiratory acidosis.

Chronic cases show bicarbonate compensation.

Analytical methods:

Blood gas analysis, capnography, and spirometry.

Specific laboratory investigations important to the study of acid-base balance and pulmonary function.

LABORATORY INVESTIGATIONS AND INTERPRET ANALYTICAL RESULTS (TABLE 2)

Table 2. Lab investigations and interpretation.

Lab investigation	*Presentation*	*Interpretation*
Alcohols (ethanol, methanol, ethylene glycol, isopropanol)	Altered mental status	Methanol and ethylene glycol ingestion causes severe acidosis, visual disturbances, or renal failure. Ethanol intoxication: euphoria, CNS depression occurs.

(Table 1) cont.....

Lab investigation	***Presentation***	***Interpretation***
Anion gap	Metabolic acidosis	Elevated anion gap indicates: organic acids (lactate, ketones), toxins (methanol, salicylate), or renal failure
Blood gas and hydrogen ion	Respiratory/metabolic acidosis or alkalosis, hypoxia	Abnormal hydrogen ion concentrations confirm severity.
Co-oximetry (carboxyhemoglobin, methemoglobin)	Carbon monoxide poisoning or methemoglobinemia	Elevated carboxyhemoglobin confirms CO poisoning Increased methemoglobin indicates oxidizing agents (*e.g.*, nitrates) Impaired oxygen transport, hypoxia.
Ketones (urine and serum)	Diabetic ketoacidosis, starvation, or alcoholic ketoacidosis	Indicates ketosis due to diabetes, starvation, or alcohol metabolism.
Lactate	Hypoxia, sepsis, shock, metabolic acidosis	Elevated lactate indicates anaerobic metabolism due to hypoxia or systemic inflammation. Patients with high lactate levels have a poor prognosis in sepsis or shock.
Osmolality	Water-electrolyte imbalances	High serum osmolality indicates dehydration Urine osmolality can differentiate renal causes of electrolyte imbalances
Osmolar Gap	Ingestion of methanol, ethylene glycol, toxins	Osmolar gap (measured osmolality – calculated osmolality) >10 indicates unmeasured osmotically active substances like alcohols or toxins.
Salicylate	Acute salicylate overdose or chronic poisoning	Elevated levels in salicylate poisoning Often associated with mixed respiratory alkalosis and metabolic acidosis Symptoms: tinnitus, hyperventilation, and altered mental status.

Summary

Acid-base and pulmonary disorders in clinical practice commonly stem from disruptions in the bicarbonate buffer system, respiratory function, or metabolic regulation. These disorders demand a thorough analysis using arterial blood gas parameters and serum electrolytes. Pulmonary function tests are crucial for evaluating respiratory mechanics and gas exchange efficiency, particularly in chronic lung diseases such as COPD and asthma. Treatment strategies hinge on identifying the primary disorder and addressing the underlying cause.

Exercises

1. Describe the compensation mechanisms for respiratory acidosis.
2. Explain the role of Pulmonary Function Tests (PFTs) in diagnosing restrictive lung diseases.
3. A patient presents with metabolic acidosis and an increased anion gap. What are the possible causes, and how would you differentiate between them?
4. How does chronic obstructive pulmonary disease lead to respiratory acidosis, and what are the potential long-term effects on the acid-base balance?

CONCLUSION

An understanding of acid-base and pulmonary function disorders is crucial for the diagnosis and management of metabolic and respiratory conditions. Clinicians require ABG results and PFTs for planning the appropriate interventions. Early detection and correction of acid-base disturbances can prevent morbidity and mortality.

ACKNOWLEDGEMENT AND CONFLICT OF INTEREST

No potential conflict of interest is declared by the author. It is also declared that the complete work is an individual effort by the author and that no financial/ administrative/ academic support was availed from any individual/ institution, or organization.

REFERENCES

[1] Achanti, A.; Szerlip, Harold, M. Acid-base disorders in the critically Ill Patient. *Clinical Journal of the American Society of Nephrology,* **2023**, (1), 102-112. [http://dx.doi.org/10.2215/CJN.04500422]

[2] Barletta, J.F.; Muir, J.; Brown, J.; Dzierba, A. A systematic approach to understanding acid-base disorders in the critically Ill. *Annals of Pharmacotherapy,* **2023**, (1), 65-75. [http://dx.doi.org/10.1177/10600280231165787]

[3] Ponce, M.C.; Sankari, A.; Sharma, S. Pulmonary function tests. In: StatPearls [Internet]. Treasure Island (FL): StatPearls Publishing; 2025 Jan. Available from: https://www.ncbi.nlm.nih.gov/books/NBK482339/

[4] Pippalapalli, J.; Lumb, A.B. The respiratory system and acid-base disorders *BJA Education,* **2023**, 221-228. [http://dx.doi.org/10.1016/j.bjae.2023.03.002]

CHAPTER 2

Cancer

Abstract: Cancer continues to stand as one of the foremost causes of morbidity and mortality worldwide, placing a substantial burden on healthcare systems. The evolution of clinical chemistry has facilitated the identification and utilization of diverse biomarkers that play pivotal roles in the early detection, diagnosis, monitoring, and therapeutic management of cancer. This module delves into the biochemical fundamentals of cancer, the significance of cancer biomarkers, the implementation of molecular diagnostic techniques, and the criticality of therapeutic monitoring. A deeper comprehension of these facets is imperative for clinical chemists to make meaningful contributions to the multidisciplinary team in the battle against cancer.

Keywords: Malignancy, paraneoplastic syndrome, ectopic hormone, multiple endocrine neoplasia, endocrine syndromes, carcinoid tumor, APUD tumors.

LEARNING OBJECTIVES

After reading this chapter, the reader should be able to understand and interpret:

- Biochemical aspects of monitoring malignant disease
- Key concepts relating to the choice, use, and measurement of biomarkers of malignancy
- Ideal tumor/cancer biomarkers
- Uses and limitations of current biomarkers of cancer
- Uses of biomarkers of cancer: prognosis, monitoring, and recurrence
- Specific laboratory investigations in the management of malignant diseases
- Analytical methods available for their measurement
- Interpret the analytical results in the context of the clinical signs and symptoms

KEY CONCEPTS

Biochemical Aspects of Monitoring Malignant Disease [1 - 5]

Monitoring of malignant diseases involves the evaluation of different biochemical factors for assessing the progression of the disease, the response to treatment, and the overall prognosis for the patient (Table 1-3).

Simmi Kharb

Table 1. Biochemical changes in cancer [3, 4].

<u>*Biochemical changes in cancer:*</u> The biochemical effects of cancer can be *direct and indirect*: *i. Direct effects*: caused by: Infiltration into normal tissue by the tumor Expansion of primary and secondary tumors *Result:* Obstruction of venous, lymphatic, or arterial supply→ Edema, infarction, retention of secretions or excretions→ Secondary infection or inflammatory response. Obstruction of the draining ducts of organs (*e.g.*, the Bile duct in the liver and the ureters from the kidneys) has profound metabolic consequences. *ii. Indirect effects:* Effects not related to direct tumor invasion of normal tissue occur due to the production of *hormones, peptides, or other biologically active factors* that act at a site remote from the tumor. Metabolic changes in cancer are not always due to aberrant hormone secretion but may be due to some other effect of the tumor or as a consequence of treatment. *Blood picture:* i. High serum calcium, uric acid, lactate, globulin, potassium, and phosphate. ii. Low serum fasting glucose, albumin, and magnesium. *iii. Anemia:* *Cause*: Hemorrhage, hemolysis, malnutrition, or bone marrow infiltration by tumor cells. iv. *Hypercoagulable state* resulting in thromboembolism 2. *Cancer cachexia:* A syndrome of weakness and generalized wasting. *Causes:* deficient food intake, intestinal obstruction or anorexia of malignancy, malabsorption, increased demands and consumption of nutrients by tumor, and the release of cachectin (TNFα).

Table 2. Tumor products as markers of malignancy.

<u>*Tumor products as markers of malignancy:*</u> Tumor products, also known as *tumor markers or biomarkers*, are substances produced by cancer cells or the body's response to cancer that can be detected in blood, urine, or tissue samples. These markers can serve as *indicators of malignancy* and provide valuable information for *cancer diagnosis, prognosis, treatment selection, monitoring, and recurrence detection.* *Common tumor products used as markers of malignancy include:* *Carcinoembryonic Antigen (CEA)* *Prostate-specific Antigen (PSA)* *CA 125* *CA 19-9* *Alpha-Fetoprotein (AFP)* *Human Chorionic Gonadotropin (hCG):* *Thyroglobulin* *HER2/neu* *Melanoma-specific Markers (e.g.*, S100, Melan-A) These tumor products play a crucial role in *cancer diagnosis, prognosis, and management,* providing valuable information to clinicians for making informed treatment decisions and optimizing patient care. However, elevated levels of these markers are not specific to cancer and can also occur in non-cancerous conditions, so their interpretation should be done in conjunction with other diagnostic tests and clinical evaluation.

Table 3. Biochemical consequences of neoplasia on other systems.

<u>*Biochemical consequences of neoplasia on other systems:*</u>
Include:
*1.**Immune System Dysregulation**:*
Tumors may *produce factors* that *suppress immune responses*:
Allowing the tumor to *evade detection and destruction by immune cells*
Tumors *may stimulate inflammatory responses*, leading to the production of cytokines and other immune mediators that can *promote tumor growth and metastasis.*
*2.**Hormonal Imbalance**:*
Certain tumors of the endocrine system secrete hormones or hormone-like substances that can disrupt normal hormonal balance
Example:
Tumors of the *adrenal glands* may produce excessive levels of *cortisol*, leading to *Cushing's syndrome*
Tumors of the pituitary gland may secrete *growth hormone*, causing *acromegaly.*
*3.**Metabolic Disturbances**:*
Tumors can alter metabolism by:
Increasing energy consumption and nutrient utilization to support their growth and proliferation
This may result in systemic metabolic disturbances:
Cachexia (muscle wasting), hyperglycemia, and dyslipidemia
*4.**Electrolyte Imbalance**:*
Some tumors of the endocrine system may *produce hormones or hormone-like substances* that *affect electrolyte balance.*
Examples:
Tumors of the *parathyroid gland* may secrete *parathyroid hormone*, leading to *hypercalcemia.*
*5.**Hematological Effects**:*
Tumors can cause *abnormalities in blood cell production* and *function*, leading to *anemia, leukopenia, and thrombocytopenia.*
Tumors may *infiltrate bone marrow, impairing blood cell production.*
*6.**Organ Dysfunction**:*
Tumors located near or within vital organs can directly *impair organ function by compressing tissues, obstructing blood flow, or causing structural damage.*
For example, *tumors in* the *brain* may *increase intracranial pressure*, leading to *neurological symptoms and deficits.*
*7.**Systemic Inflammation**:*
Chronic inflammation, being a *hallmark of cancer*, can contribute to tumor progression and metastasis.
Tumors produce inflammatory cytokines and chemokines that stimulate immune cell infiltration and promote angiogenesis.
*8.**Oxidative Stress**:*
Reactive Oxygen Species (ROS) are produced by tumors, leading to oxidative stress and cellular damage.
Neoplasia causes widespread biochemical effects on the body, affects multiple systems, and contributes to the *systemic manifestations of cancer.*
Understanding of these biochemical consequences is essential for developing targeted therapies and interventions for managing cancer-related complications and improve patient outcomes.

KEY BIOCHEMICAL ASPECTS COMMONLY MONITORED IN MALIGNANT DISEASES: [3, 4]

Tumor Markers

Definition

Tumor markers are substances produced by cancer cells or the body in response to cancer.

Application

They can be used for screening for cancer, diagnosing specific types of cancer, monitoring treatment response, and detecting cancer recurrence.

These markers can be detected in blood, urine, or tissue samples.

Examples:

Prostate-Specific Antigen (PSA) for prostate cancer, CA 125 for ovarian cancer, and CEA (Carcinoembryonic Antigen) for colorectal cancer.

Genetic Biomarkers

Definition:

Genetic biomarkers are changes in the DNA or RNA of cancer cells that can predict disease prognosis, guide treatment decisions, and monitor response to treatment.

Techniques:

Next-Generation Sequencing (NGS) and Polymerase Chain Reaction (PCR) are capable of identifying genetic mutations, gene amplifications, and chromosomal rearrangements linked to malignant diseases.

Metabolic Markers

Metabolic markers are indicative of alterations in cellular metabolism that are associated with the growth and proliferation of cancer.

Techniques

Metabolic imaging techniques:

Positron Emission Tomography (PET) and *Magnetic Resonance Spectroscopy* (MRS) can assess glucose metabolism, lactate production, and other metabolic processes within tumors. The P*ET metabolic probe [^{18}F] Fluoro-Deoxyglucose ([^{18}F]FDG)* is the most widely used because it accumulates in highly glycolytic tissues.

Application

Monitoring metabolic markers can offer valuable information about tumour aggressiveness, response to therapy, and disease recurrence.

Inflammatory Markers

In cancer patients, inflammatory markers such as *C-Reactive Protein (CRP), Interleukin-6 (IL-6), and Tumor Necrosis Factor-alpha (TNF-alpha)* are elevated due to chronic inflammation in the tumor microenvironment.

Application

Monitoring inflammatory markers can help assess disease activity, predict treatment response, and identify patients at higher risk of cancer-related complications.

Hormonal Markers

Hormonal markers are hormones or hormone receptors that actively contribute to the growth and progression of hormone-sensitive cancers, including *breast cancer and prostate cancer*.

Monitoring hormone levels, such as estrogen, progesterone, and testosterone, can help guide hormone-based therapies and assess treatment efficacy in hormone-dependent malignancies.

Drug Levels and Metabolites

Monitoring the levels of chemotherapy drugs and their metabolites in the blood or urine is crucial for assessing drug exposure, metabolism, and elimination kinetics.

Therapeutic Drug Monitoring (TDM) helps optimize chemotherapy dosing, minimize toxicity, and ensure adequate drug efficacy in cancer patients undergoing chemotherapy treatment.

Immune Markers

Immune markers, such as immune cells, cytokines, and checkpoint proteins, are crucial components in the body's response to cancer.

Various immunotherapy treatments, namely immune checkpoint inhibitors and targeted immune markers, enhance anti-tumor immunity and control cancer growth. Monitoring immune markers can help predict response to immunotherapy and identify patients likely to benefit from immune-based treatments.

Oncogenic Signaling Pathways

Aberrant activation of oncogenic signaling pathways, such as the PI3K/AKT/mTOR and MAPK/ERK pathways, can lead to cancer cell proliferation, survival, and metastasis.

Targeted therapies have been developed to treat cancers with activating mutations by inhibiting specific signalling molecules or pathways. It is crucial to monitor oncogenic signalling pathways to guide the selection of targeted therapies and assess treatment response in patients with molecularly defined cancers.

Monitoring malignant diseases requires assessing various biochemical markers, including tumor markers, genetic biomarkers, metabolic markers, inflammatory markers, hormonal markers, drug levels, immune markers, and oncogenic signalling pathways. Integrating these biochemical assessments into clinical practice empowers clinicians to tailor treatment strategies, monitor disease progression, and improve patient outcomes in cancer management.

KEY CONCEPTS RELATING TO CHOICE, USE, AND MEASUREMENT OF BIOMARKERS OF MALIGNANCY [3, 4, 6]

The choice, use, and measurement of biomarkers of malignancy are critical aspects of cancer diagnosis, prognosis, treatment selection, and monitoring.

These key concepts provide a framework for understanding the role of biomarkers in malignancy diagnosis, prognosis, treatment, and monitoring, highlighting the importance of thoughtful biomarker selection, validation, interpretation, and integration into clinical practice (Table **4**).

Table 4. Biomarkers in malignancy.

Biomarker	*Choice*	*Use*	*Measurement*
Alpha-fetoprotein (AFP)	Liver cancer diagnosis, especially hepatocellular carcinoma. Also used in germ cell tumors and yolk sac tumors.	Liver cancer screening, monitoring treatment response, and detecting recurrence.	Blood test (serum AFP levels).
Carcinoembryonic Antigen (CEA)	Colorectal cancer, lung, breast, and pancreatic cancer monitoring and prognosis.	To monitor treatment response, detect recurrence, and assess prognosis.	Blood test (serum CEA levels).
CA 125	Biomarker for ovarian cancer in monitoring treatment response and detecting recurrence.	Monitoring of ovarian cancer treatment response, assessing prognosis, and detecting recurrence.	Blood test (serum CA 125 levels).
CA 19-9	Pancreatic and gastrointestinal cancer diagnosis, prognosis, and monitoring.	Pancreatic cancer diagnosis, monitoring treatment response, and detecting recurrence.	Blood test (serum CA 19-9 levels).
Human Chorionic Gonadotropin (hCG)	Gestational trophoblastic diseases, including choriocarcinoma and testicular germ cell tumors.	Gestational trophoblastic disease diagnosis, monitoring treatment response, and detecting recurrence.	Blood or urine test (serum or urine hCG levels).
HER2/neu	Biomarker of breast cancer, guiding HER2-targeted therapies.	HER2-positive breast cancer identification, predicting response to HER2-targeted therapies, and assessing prognosis.	Immunohistochemistry (IHC) or Fluorescence *In Situ* Hybridization (FISH) to evaluate HER2 protein expression or gene amplification.
Melanoma-specific Markers (*e.g.*, S100, Melan-A)	Melanoma diagnosis, distinguishing melanoma from other skin lesions.	Melanoma diagnosis, monitoring treatment response, and detecting recurrence.	Immunohistochemistry (IHC) staining of tissue samples.
Prostate-Specific Antigen (PSA)	Prostate cancer screening, diagnosis, and monitoring.	Screening for prostate cancer, monitoring treatment response, and detecting recurrence.	Blood test (serum PSA levels).
Thyroglobulin	Biomarker of thyroid cancer, in monitoring recurrence and assessing treatment response after thyroidectomy.	Monitoring of thyroid cancer treatment response, detection of recurrence, and assessing prognosis.	Blood test (serum thyroglobulin levels).

KEY CONCEPTS RELATING TO THE SELECTION, UTILIZATION, AND MEASUREMENT OF BIOMARKERS IN MALIGNANCY

Ideal Tumor/Cancer Biomarkers

Ideal biochemical tumor/cancer biomarkers possess several key characteristics that make them valuable for cancer detection, diagnosis, prognosis, treatment selection, and monitoring (Table 5). Desirable qualities of ideal tumor/cancer biomarkers include:

Table 5. Ideal tumor markers.

Tumor Marker Quality	*Benefits*
High sensitivity and specificity	Ensure reliable detection Minimize false-positive and false-negative results
Early detection	Facilitate timely intervention Improve patient survival rates
Clinical validity and utility	Provide information to guide clinical decision-making Improves patient outcomes
Prognostic value	Provide information about the likely course and outcome of cancer, enabling personalized patient management
Predictive value	Serve as predictive indicators of response to identify patients who are most likely to benefit from a particular treatment approach
Non-invasive detection	Detectable through minimally invasive or non-invasive methods, such as *blood tests, urine tests, or imaging techniques.* Convenient, cost-effective, and well-tolerated by patients, facilitating widespread screening and monitoring
Dynamic monitoring	Provide real-time information about changes in tumor burden, disease activity, and treatment efficacy Enable timely adjustments to treatment regimens and disease management strategies
High Stability and Reproducibility in sample collection, processing, and measurement	Robust and reliable across different laboratory settings Ensures consistent and accurate results in clinical practice
Cost-effective	Allow widespread adoption in routine clinical practice and population-based screening programs Cost-effective biomarkers minimize healthcare costs while maximising the benefits of early detection and intervention
Adhere to Ethical Considerations in research and clinical use	Ensures that patient rights are respected and protected throughout the diagnostic and therapeutic process

Ideal biochemical tumor/cancer biomarkers must exhibit high sensitivity and specificity, enable early detection, demonstrate clinical validity and utility, provide prognostic and predictive value, offer non-invasive detection, allow for dynamic monitoring, exhibit high stability and reproducibility, be cost-effective, and adhere to ethical principles.

Integrating ideal biomarkers into cancer care pathways enhances precision medicine approaches, improves patient outcomes, and advances the field of oncology.

USES AND LIMITATIONS OF CURRENT BIOMARKERS OF CANCER

Biomarkers of cancer play diverse roles in cancer management, including prognostication, monitoring treatment response, and detecting recurrence (Table 6) [6, 7].

Table 6. Biomarkers of cancer.

Biomarker	*Uses: Screening, Diagnosis, and Donitoring of:*	*Limitations*
Alpha-Fetoprotein (AFP)	Hepatocellular carcinoma	- Limited sensitivity, particularly for early-stage liver cancer detection. - *Can be elevated* in non-cancerous liver diseases: *Cirrhosis, Hepatitis* *-Not all liver cancers produce AFP* *Other cancers may also cause AFP elevations*, limiting its specificity.
Carcinoembryonic Antigen (CEA)	- Colorectal, lung, breast, and pancreatic cancers.	*-Lacks specificity*: *Levels are elevated:* inflammation, infection, and smoking *-Limited sensitivity*: especially for early-stage cancers. - Levels may not *correlate* with *tumor burden or disease progression* in all cases.
CA 125	- Ovarian cancer	*-Lack of sensitivity* for early-stage ovarian cancer detection *-Elevated in non-cancerous conditions:* non-ovarian cancers, endometriosis, and Pelvic Inflammatory Disease (PID) *-Limited specificity*: not all ovarian cancers produce elevated CA 125 levels
CA 19-9	- Pancreatic cancer patients Assessing prognosis in pancreatic cancer	*-Limited sensitivity and specificity*: elevated levels can occur in benign pancreatic conditions and other malignancies. - Levels *may not correlate with tumor stage* or disease progression in all cases

(Table 6) cont.....

Human Chorionic Gonadotropin (hCG)	- Gestational trophoblastic diseases: choriocarcinoma and hydatidiform mole - Detecting recurrence	*-Limited specificity: elevated levels* can occur in: normal pregnancy, ectopic pregnancy, and certain non-trophoblastic tumours - Conditions such as *testicular germ cell tumors may also produce elevated hCG levels: false-positive* results in non-gestational trophoblastic diseases.
HER2/neu	- HER2-positive breast cancer	*-Limited utility in cancers other than breast cancer*, as HER2 overexpression is less common in other malignancies. - HER2 testing may be subject to *inter-laboratory variability and discordance between different testing methods* (*e.g.*, immunohistochemistry, fluorescence *in situ* hybridization).
Melanoma-specific Markers (*e.g.*, S100, Melan-A)	- Melanoma and distinguishing it from benign skin lesions.	*-Limited sensitivity*: not all melanomas produce elevated levels of these markers. *-Limited specificity:* This may also be *expressed in benign conditions and other malignancies*
Prostate-specific Antigen (PSA)	- Prostate cancer	*-Limited specificity*: elevated levels can occur in benign prostatic conditions: prostatitis and Benign Prostatic Hyperplasia (BPH). *-Overdiagnosis and overtreatment* of indolent prostate cancers due to false-positive results. - PSA levels may not accurately distinguish between aggressive and non-aggressive prostate cancers.
Thyroglobulin	- Thyroid cancer	*-Limited specificity: elevated levels can occur in non-cancerous thyroid conditions*: thyroiditis and goitre. - Thyroglobulin levels *may not accurately reflect disease status* in patients who have undergone *thyroidectomy and are on thyroid hormone replacement therapy.*

USES OF BIOMARKERS OF CANCER: PROGNOSIS, MONITORING, AND RECURRENCE

Biomarkers of cancer play diverse roles in cancer management, including prognostication, monitoring treatment response, and detecting recurrence (Table 7).

Table 7. Uses of biomarkers of cancer.

Uses	Description
Prognosis	Biomarkers are used to predict the likely outcome of cancer, including survival rates, disease progression, and response to treatment, helping clinicians tailor patient management.
Monitoring	Biomarkers are used to monitor cancer progression, treatment response, and recurrence, providing valuable insights into disease dynamics and guiding therapy adjustments.

In summary, cancer biomarkers are versatile tools used for prognosis, monitoring treatment response, detection of recurrence, guiding treatment decisions, and advancing research and drug development efforts. Their multifaceted roles contribute to improved patient outcomes, personalized treatment approaches, and advancements in cancer care [6 - 8].

SPECIFIC LABORATORY INVESTIGATIONS IN THE MANAGEMENT OF MALIGNANT DISEASES

Laboratory investigations play a crucial role in the management of malignant diseases, providing valuable information for diagnosis, prognosis, treatment selection, and monitoring (Table **8**) [6 - 8].

Table 8. Commonly used specific laboratory investigations in the management of malignant diseases.

Laboratory Investigations	*Description*
Tumor Markers	Measurement of *specific biomarkers in blood or tissue samples* *Examples:* Prostate-Specific Antigen (PSA) for prostate cancer CA 125 for ovarian cancer CEA for colorectal cancer
Genetic Testing	*Analysis of genetic mutations or alterations* associated with cancer risk, prognosis, and treatment response *Example:* Testing for BRCA1/2 mutations in breast and ovarian cancer
Histopathological Examination	*Microscopic examination of tissue* samples to assess the morphology and characteristics of cancer cells for diagnosis, staging, and classification of malignancies
Immunohistochemistry	*Detection of specific proteins or antigens* in tissue samples using antibodies to gather additional information about tumor markers and subtypes for targeted therapy selection
Cytogenetic Analysis	*Evaluation of chromosomal abnormalities and gene rearrangements* in cancer cells, assisting in diagnosis, prognosis, and identifying therapeutic targets in *hematological malignancies*
Flow Cytometry	*Quantitative analysis of cell populations* based on their physical and chemical characteristics, useful for *assessing tumor cell markers, proliferation, and immunophenotyping*
Molecular Profiling	*Analysis of genetic alterations, gene expression patterns, and molecular pathways* in cancer cells, guiding personalized treatment strategies and predicting response to targeted therapies

LABORATORY INVESTIGATIONS IN THE MANAGEMENT OF MALIGNANT DISEASES

Specific laboratory investigations in the management of malignant diseases encompass a wide range of tests and techniques, including CBC, peripheral blood smear, serologic tests, biochemical tests, tumor markers, immunohistochemistry, molecular genetic testing, flow cytometry, cytogenetic studies, and NGS [1, 6, 7].

These laboratory investigations contribute to cancer diagnosis, prognosis, treatment selection, and monitoring, guiding personalized patient care and improving clinical outcomes (Table **9**).

Table 9. Laboratory investigations in the management of malignant diseases.

Laboratory Investigation	*Description/ application*
Complete Blood Count (CBC)	Levels of various blood cells in underlying malignancies: Leukemia or lymphoma Effects of cancer treatment: Chemotherapy-induced cytopenia
Peripheral Blood Smear	Microscopic examination of blood cells: Identification of abnormal cell morphology: *Immature or atypical cells* characteristic of certain leukemias and lymphomas
Serologic Tests	To detect antibodies or antigens in blood serum To provide diagnostic or prognostic information for certain malignancies *E.g.*, hepatitis B or C virus infection-associated hepatocellular carcinoma or Human Papillomavirus (HPV) infection-associated cervical cancer
Biochemical Tests	*Liver Function Tests (LFTs):* To assess liver function To detect liver damage or dysfunction: Primary liver cancer (hepatocellular carcinoma) Metastatic disease involving the liver *Renal Function Tests:* To evaluate kidney function: Assess kidney damage due to: Cancer-related complications or Nephrotoxic effects of chemotherapy drugs *Electrolyte Levels:* Serum sodium, potassium, and calcium To monitor electrolyte imbalances associated with malignancies, dehydration, or treatment-related toxicities

(Table 9) cont.....

Laboratory Investigation	*Description/ application*
Tumor Markers	Aids in: Cancer diagnosis Monitoring treatment response Detecting disease recurrence: *E.g.*, Prostate-Specific Antigen (PSA) for prostate cancer CA 125 for ovarian cancer CEA (carcinoembryonic antigen) for colorectal cancer
Immunohistochemistry (IHC)	To detect specific proteins or antigens in tissue samples obtained from biopsies or surgical specimens: *E.g.*, Hormone receptors or HER2/neu in breast cancer
Molecular Genetic Testing	*Fluorescence In Situ* Hybridization (FISH): Detects specific chromosomal abnormalities or gene rearrangements associated with certain malignancies: HER2 amplification in breast cancer BCR-ABL translocation in Chronic Myeloid Leukemia (CML) *Polymerase Chain Reaction (PCR):* Amplifies and detects specific DNA sequences Identification of genetic mutations, gene expression patterns relevant to cancer diagnosis, prognosis, and treatment selection
Flow Cytometry	Analyzes characteristics of individual cells in suspension: Cell size, shape, and surface marker expression. Used to diagnose and classify hematologic malignancies, *e.g.*, leukemia and lymphoma, based on the immunophenotypic profile of abnormal cells
Cytogenetic Studies	*Karyotyping or Comparative Genomic Hybridization (CGH):* To examine chromosomal abnormalities or karyotypes of cancer cells *Cytogenetic analysis:* To identify associated recurrent chromosomal aberrations and genetic alterations in specific malignancies Aids in diagnosis, prognosis, and treatment planning
Next-Generation Sequencing (NGS)	NGS *enables comprehensive analysis of* cancer genomes, identifying mutations, copy number alterations, and gene expression profiles relevant to cancer development and progression. Used to identify actionable mutations, predict treatment response, and guide targeted therapy in precision oncology.

LAB INVESTIGATIONS FOR MANAGEMENT OF MALIGNANT DISEASES

Analytical Methods Available for the Measurement of Biomarkers of Cancer

Analytical methods are available for the measurement of biomarkers of cancer, and they offer specific advantages in terms of sensitivity, specificity, throughput, and cost-effectiveness.

Conventional immunochemical techniques for the measurement of target markers still dominate the area of laboratory investigation and clinical application.

However, in addition to radioimmunoassay and enzyme-linked immunosorbent techniques, immunohistochemistry, flow cytometry, imaging techniques, and new promising molecular biology procedures are available (Table **10**).

Table 10. Analytical methods for the measurement of biomarkers of cancer [1-5].

Analytical Method	*Principle*	*Applications*
Immunoassays	Encompass a broad range of techniques: Radioimmunoassay (RIA), Enzyme Immunoassay (EIA), Chemiluminescent Immunoassay (CLIA) These methods are based on the specific binding of antibodies to target biomarkers, followed by detection of the antibody-antigen complex using various detection methods, such as radioactivity, enzyme activity, or luminescence.	Immunoassays offer high sensitivity and specificity and are adaptable to high-throughput screening applications.
Enzyme-Linked Immunosorbent Assay (ELISA)	Quantitative detection of specific antigens or antibodies using enzyme-labelled antibodies conjugated to enzymes Detector: colorimetric or fluorometric detection.	Highly sensitive, specific, and relatively easy to perform, suitable for clinical biomarker analysis.
Polymerase Chain Reaction (PCR)	Amplification of specific DNA sequences in patient samples Detection and quantification of genetic mutations, gene expression, and viral DNA in cancer cells.	
Immunohistochemistry (IHC)	Visualization of specific proteins or antigens in tissue samples using labelled antibodies and microscopy Provides spatial and qualitative information about biomarkers.	
Mass Spectrometry	Based on the mass-to-charge ratio of proteins and peptides. Liquid Chromatography-Tandem Mass Spectrometry (LC-MS/MS) and Matrix-Assisted Laser Desorption/Ionization Mass Spectrometry (MALDI-MS) are commonly used MS-based methods. Enabling high-throughput biomarker analysis. Identification and quantification of proteins, peptides, and metabolites in patient samples.	Offers high sensitivity, specificity, and multiplexing capabilities, allowing for the simultaneous measurement of multiple biomarkers in complex biological samples, for biomarker discovery and validation in cancer research.

(Table 10) cont.....

Analytical Method	*Principle*	*Applications*
Flow Cytometry	Quantitative analysis of cell populations based on their physical and chemical properties: their size, granularity, and surface marker expression. Fluorescently labelled antibodies or fluorescent dyes are used to detect and quantify biomarkers on the cell surface or within cells. Detection and enumeration of tumour-specific markers on individual cells.	Commonly used in hematologic malignancies to characterize cell populations and detect aberrant immunophenotypes associated with cancer.
Next-Generation Sequencing (NGS)	High-throughput DNA sequencing technology. Gives a comprehensive analysis of genetic mutations, copy number variations, and gene expression profiles in cancer cells.	Commonly used NGS approaches in cancer biomarker research and clinical diagnostics.
Liquid Biopsy	Analysis of *Circulating Tumor Cells (CTCs), cell-free DNA (cfDNA), extracellular vesicles (exosomes),* and *other biomarkers* in blood samples, providing *minimally invasive cancer monitoring.*	
Immunofluorescence (IF)	Uses fluorescently labelled antibodies and microscopy Enables sensitive and multiplexed biomarker detection of specific proteins or antigens in tissue samples.	
Imaging Techniques	Positron Emission Tomography (PET), Computed Tomography (CT), Magnetic Resonance Imaging (MRI), ultrasound, radiolabelled tracers targeting specific biomarkers, such as glucose metabolism (FDG-PET) or Prostate-Specific Membrane Antigen (PSMA-PET).	PET, CT, MRI, and ultrasound can indirectly assess cancer biomarkers by visualizing tumor morphology, metabolism, or function. FDG-PET and PSMA-PET are used in molecular imaging for cancer diagnosis, staging, and assessment of treatment response.
Microarray	Enable the simultaneous analysis of thousands of genes or proteins in a single experiment. Types: DNA microarrays Protein microarrays	Profile gene expression patterns or protein biomarker levels in cancer samples. Provide insights into cancer biology, molecular subtypes, and patient stratification.

ANALYTICAL METHODS USED FOR MEASURING BIOMARKERS OF CANCER

These analytical methods offer diverse capabilities for measuring biomarkers of cancer, providing valuable information for cancer diagnosis, prognosis, treatment selection, and monitoring. Depending on factors such as the type of biomarker, sample type, required sensitivity and specificity, throughput, and available resources, the most appropriate analytical method can be selected.

Integration of multiple analytical approaches can comprehensively characterize cancer biomarkers and their clinical significance.

INTERPRET THE ANALYTICAL RESULTS IN THE CONTEXT OF THE CLINICAL SIGNS AND SYMPTOMS [6 - 8]

AFP

AFP Levels

- *Normal blood AFP levels*: less than 10 ng/mL in adults.
- *Elevated AFP levels:* can be present in:

Hepatocellular Carcinoma (HCC)

AFP is often elevated in patients with HCC, but not in all cases of HCC.

Persistent AFP elevation or rising AFP levels in a patient with liver disease may indicate the presence of HCC.

Diagnostic evaluations such as imaging studies (Ultrasound, CT Scan, MRI) and clinical findings should be considered to confirm the diagnosis and staging of HCC and other liver tumors.

Germ Cell Tumors

Elevated AFP levels are seen in certain germ cell tumors

E.g., non-seminomatous testicular cancer and ovarian cancer.

Often used in conjunction with other tumor markers, *e.g.*, beta-human chorionic gonadotropin (β-hCG), for the diagnosis and monitoring of these tumors.

- Serial monitoring of AFP levels over time can provide valuable information about disease progression or response to treatment.
- Low AFP levels: usually not clinically significant.

Interpretation of AFP results requires careful consideration of clinical presentation, underlying conditions, imaging findings, and trend analysis of AFP levels over time. AFP elevation alone is not diagnostic of cancer but serves as a valuable adjunctive tool in the evaluation and management of patients with suspected or known liver tumors or germ cell tumors.

ALP Isoenzymes

Occurrence

Alkaline Phosphatase (ALP) enzyme is found in various tissues throughout the body, including the liver, bone, kidney, and intestine.

ALP isoenzyme analysis can help differentiate the tissue source of elevated ALP levels, providing valuable information in the context of clinical signs and symptoms.

Total ALP Level

Elevated total ALP levels may indicate increased enzyme activity originating from various tissues.

However, total ALP alone does not provide information about the specific tissue source of the elevation.

ALP Isoenzymes

Clinical signs and symptoms associated with specific diseases or conditions should be correlated with the tissue source of elevated ALP levels.

Liver ALP (ALP-L)

Elevated liver ALP levels are commonly seen in liver diseases, such as *hepatitis, cholestasis, cirrhosis, or liver metastases.*

Clinical signs and symptoms associated with *liver disease include jaundice, abdominal pain, hepatomegaly, pruritus, dark urine, and pale stools.*

Abnormal liver function tests (e.g., elevated bilirubin and transaminases).

Bone ALP (ALP-B)

Elevated bone ALP levels are often observed in conditions involving increased bone turnover, such as *Paget's disease, osteomalacia, osteoporosis, or bone metastases.*

Clinical signs and symptoms may include *bone pain, fractures, bone deformities, height loss,* or *elevated serum calcium or phosphate levels.*

Intestinal ALP (ALP-I)

Elevated intestinal ALP levels can occur in intestinal disorders, such as *Inflammatory Bowel Disease (IBD), celiac disease, or intestinal obstruction.*

Clinical manifestations may include *diarrhea, abdominal pain, malabsorption,* bloating, *or weight loss.*

Imaging Studies

Imaging modalities, such as *ultrasound, CT scan, MRI, or bone scans*, may help identify structural abnormalities or lesions associated with liver, bone, or intestinal diseases. Imaging findings, in conjunction with ALP isoenzyme analysis, aid in the diagnosis and management of underlying conditions.

Additional Laboratory Tests

Other laboratory tests, including liver function tests (*e.g., bilirubin, AST, ALT), bone markers (e.g.,* calcium, phosphate), or specific serologic tests (*e.g.*, autoimmune markers, tumor markers), may provide further insights into diagnosis and help confirm the underlying etiology of elevated ALP levels.

Interpretation of ALP isoenzyme results involves correlating tissue-specific ALP elevations with the clinical signs, symptoms, and diagnostic findings associated with liver, bone, or intestinal diseases.

A comprehensive assessment, including clinical evaluation, imaging studies, and additional laboratory tests, is essential for accurate diagnosis and appropriate management of patients with elevated ALP levels.

CA 15-3

Importance:

CA 15-3 is a tumor marker primarily associated with *breast cancer*, and it can also be elevated in other malignancies and non-cancerous conditions.

Normal Reference Range:

CA 15-3 levels: normal reference range less than 30 U/mL.

Elevated CA 15-3 levels may suggest an underlying pathology.

Elevated CA 15-3 levels:

Breast Cancer:

Most commonly monitored in breast cancer patients.

Elevated CA 15-3 levels may indicate active disease, disease progression, or recurrence.

Clinical signs and symptoms of breast cancer include *breast lumps, changes in breast size or shape, nipple discharge, skin changes, and lymph node enlargement.*

Other Cancers:

Elevated CA 15-3 levels may also be observed in other cancers, including *ovarian, lung, pancreatic, and gastrointestinal cancers.*

Clinical signs and symptoms associated with these malignancies should be considered in conjunction with CA 15-3 levels for diagnostic evaluation.

Non-cancerous Conditions:

CA 15-3 levels can be elevated in non-cancerous conditions such as *liver disease, benign breast conditions (e.g.*, fibrocystic changes), inflammatory conditions, and infections.

Clinical history and physical examination findings help differentiate between benign and malignant causes of elevated CA 15-3 levels.

Elevated CA 15-3 levels should be interpreted in conjunction with other *clinical and radiological assessments (mammography, ultrasound, MRI, CT scan, or PET scan), additional diagnostic tests (Biopsy and histopathological examination of tissue samples)*, along with *other tumor markers (e.g.*, CA 27-29, CEA), complete blood count, liver function tests, as changes in CA 15-3 levels alone may not always accurately reflect disease status.

Interpretation of CA 15-3 results requires consideration of the patient's clinical history, symptoms, imaging findings, and other laboratory tests. Elevated CA 15-3 levels may indicate breast cancer or other malignancies, but clinical correlation and further diagnostic evaluation are necessary for accurate diagnosis and management.

CA 19-9

CA 19-9 is a tumor marker primarily associated with gastrointestinal malignancies, particularly pancreatic cancer.

Normal Reference Range

CA 19-9 levels: typically less than 37 U/mL.

Elevated CA 19-9 levels may suggest an underlying pathology.

Pancreatic Cancer

CA 19-9 levels are most commonly monitored in *pancreatic cancer patients.*

Elevated CA 19-9 levels may indicate pancreatic cancer, particularly in conjunction with clinical signs and symptoms such as jaundice, abdominal pain, weight loss, loss of appetite, nausea, vomiting, and changes in bowel habits.

Other Gastrointestinal Cancers

Elevated CA 19-9 levels can also be observed in other gastrointestinal malignancies, including *bile duct cancer, gastric cancer, colorectal cancer, and hepatobiliary cancer.*

Clinical signs and symptoms specific to these cancers should be considered in conjunction with CA 19-9 levels for diagnostic evaluation.

Benign Conditions

CA 19-9 levels can be elevated in *non-cancerous conditions such as pancreatitis, biliary obstruction, cholangitis, liver cirrhosis, and inflammatory bowel disease.*

Clinical history and physical examination findings *help differentiate between benign and malignant causes* of elevated CA 19-9 levels.

Rising CA 19-9 levels may indicate *tumor growth or recurrence*

Decreasing CA 19-9 levels may suggest a favorable response to therapy.

CA 19-9 levels should be interpreted in conjunction with other *clinical and radiological assessments (CT scan, MRI, Endoscopic Ultrasound (EUS), or PET scan), biopsy and histopathological examination of tissue samples, additional laboratory tests, including other tumor markers (e.g.*, CEA, CA 125), complete blood count, liver function tests, t*o* help guide diagnosis, staging, and treatment decisions, as changes in CA 19-9 levels alone may not always accurately reflect disease status.

Interpretation of CA 19-9 results requires consideration of the patient's clinical history, symptoms, imaging findings, and other laboratory tests. Elevated CA 19-

9 levels may indicate pancreatic cancer or other gastrointestinal malignancies, but clinical correlation and further diagnostic evaluation are necessary for accurate diagnosis and management.

CA 125

CA 125 is a tumor marker primarily associated with ovarian cancer, although it can also be elevated in other gynecological malignancies and non-cancerous conditions:

Normal Reference Range: usually less than 35 U/mL

Elevated CA 125 levels:

Ovarian Cancer

Most commonly used for monitoring disease progression and treatment response in ovarian cancer patients.

Elevated CA 125 levels may indicate *ovarian cancer*, particularly in conjunction with clinical signs and symptoms such as abdominal or pelvic pain, bloating, increased abdominal girth, urinary urgency or frequency, changes in bowel habits, and unexplained weight loss.

Other Gynecological Malignancies

Elevated CA 125 levels can also be observed in other gynecological cancers, including *endometrial cancer, fallopian tube cancer, and peritoneal cancer.*

Benign Gynecological Conditions

CA 125 levels can be elevated in *non-cancerous conditions such as endometriosis, ovarian cysts, Pelvic Inflammatory Disease (PID), uterine fibroids, and menstruation.*

Interpretation of CA 125 results requires consideration of the patient's clinical history, symptoms, imaging findings, and other laboratory tests.

CA 27-29

CA 27-29 is a tumor marker primarily associated with breast cancer, particularly in patients with advanced or metastatic disease.

Normal Reference Range

Usually less than 38 U/mL.

Elevated CA 27-29 levels:

Breast Cancer

Elevated CA 27-29 levels may indicate breast cancer, particularly in conjunction with clinical signs and symptoms such as *breast lumps, changes in breast size or shape, nipple discharge, skin changes, and lymph node enlargement.*

Monitoring disease progression and response to treatment.

Interpretation of CA 27-29 results requires consideration of the patient's clinical history, symptoms, imaging findings, and other laboratory tests.

Calcitonin

Calcitonin hormone is primarily produced by the thyroid gland. Interpretation of calcitonin results in the context of clinical signs and symptoms involves several considerations, particularly in the assessment of thyroid disorders and other conditions. Imaging studies such as *thyroid ultrasound, Fine-Needle Aspiration (FNA) biopsy, and genetic testing for mutations in the RET proto-oncogene*

Normal Reference Range

Fasting calcitonin levels in healthy individuals are less than 10 pg/mL.

Elevated levels:

Medullary Thyroid Carcinoma (MTC)

Elevated calcitonin levels are most commonly observed. MTC is a rare, aggressive form of thyroid cancer arising from the parafollicular or C cells of the thyroid gland.

Clinical signs and symptoms of MTC may include a *palpable thyroid nodule or mass, hoarseness, dysphagia (difficulty swallowing), neck pain, lymph node enlargement, and symptoms of hypercalcemia if metastasis is present.*

Thyroid Nodules

Calcitonin levels may be elevated in *benign thyroid conditions such as C-cell hyperplasia or thyroid nodules*, particularly in cases of sporadic hypercalcitoninemia.

Interpretation of calcitonin results involves careful consideration of the patient's clinical signs and symptoms, along with diagnostic evaluation, including imaging studies, biopsy, and genetic testing, to differentiate benign from malignant causes of elevated calcitonin levels, particularly in the context of suspected medullary thyroid carcinoma.

CEA

Carcinoembryonic Antigen (CEA) is a tumor marker commonly associated with colorectal cancer, although elevated levels can be seen in various other malignancies and non-cancerous conditions, and serial monitoring of CEA levels is often used to assess disease progression or response to treatment

Normal Reference Range

Normal CEA levels are less than 3 to 5 ng/mL in non-smokers and less than 5 to 10 ng/mL in smokers.

Elevated levels

Colorectal Cancer

Elevated CEA levels may indicate colorectal cancer, particularly in conjunction with clinical signs and symptoms such as *rectal bleeding, changes in bowel habits (e.g.*, diarrhea or constipation), abdominal pain, unexplained weight loss, anemia, or palpable abdominal masses.

Other Cancers

Elevated CEA levels can also be observed in other malignancies, including *lung cancer, pancreatic cancer, breast cancer, gastric cancer, and liver cancer.*

Clinical signs and symptoms specific to these cancers should be considered in conjunction with CEA levels for diagnostic evaluation.

Non-cancerous Conditions

CEA levels can be elevated in non-cancerous conditions such as *inflammatory bowel disease (e.g.*, Crohn's disease, ulcerative colitis), pancreatitis, liver cirrhosis, Chronic Obstructive Pulmonary Disease (COPD), and smoking.

Interpretation of CEA results requires consideration of clinical history, symptoms, imaging findings, and other laboratory tests.

Elevated CEA levels may indicate colorectal cancer or other malignancies, but clinical correlation and further diagnostic evaluation are necessary for accurate diagnosis and management.

HE4

Human Epididymis Protein 4 (HE4) is a tumor marker primarily associated with ovarian cancer, particularly epithelial ovarian carcinoma.

Normal Reference Range

Normal HE4 levels are less than 140 pmol/L.

Elevated HE4

Ovarian Cancer

Elevated HE4 levels may indicate *ovarian cancer*, particularly in conjunction with clinical signs and symptoms such as *abdominal or pelvic pain, bloating, increased abdominal girth, urinary urgency or frequency, changes in bowel habits, and unexplained weight loss.*

Interpretation of HE4 results requires consideration of clinical history, symptoms, imaging findings, and other laboratory tests.

Elevated HE4 levels may indicate ovarian cancer, particularly in the context of clinical signs and symptoms suggestive of malignancy, but further diagnostic evaluation is necessary for accurate diagnosis and management.

hCG

Human Chorionic Gonadotropin (hCG) is a hormone produced by the placenta during pregnancy.

Normal reference range

In non-pregnant individuals, hCG levels are typically very low, usually less than 5 mIU/mL.

In pregnant women, hCG levels rise rapidly in early pregnancy, doubling approximately every 48 to 72 hours during the first

Elevated levels

<u>Pregnancy confirmation and monitoring</u>

Elevated hCG levels are typically indicative of *pregnancy.*

Elevated hCG levels (*e.g., trophoblastic disease, molar pregnancy or choriocarcinoma, ectopic pregnancy, and certain medications).*

Additional laboratory tests and imaging studies may be indicated based on clinical suspicion and hCG levels to evaluate for potential pregnancy-related complications.

A failure of hCG levels to rise appropriately or a decline in hCG levels may indicate a potential *miscarriage or ectopic pregnancy.*

Clinical signs and symptoms may include *abnormal vaginal bleeding, enlarged uterus, and signs of hyperthyroidism.*

Interpretation of hCG results involves consideration of the patient's clinical signs and symptoms, gestational age (if applicable), and clinical context. Elevated hCG levels are typically indicative of pregnancy, but further diagnostic evaluation may be necessary to assess for potential pregnancy-related complications or trophoblastic disease in certain clinical scenarios.

HER2/Neu

Human Epididymis protein 4 (HE4) is a tumor marker primarily associated with ovarian cancer, particularly epithelial ovarian carcinoma.

Normal range for HE4 levels is less than 140 pmol/L.

Elevated levels

<u>Ovarian Cancer</u>

Elevated HE4 levels may indicate ovarian cancer, particularly in conjunction with clinical signs and symptoms such as *abdominal or pelvic pain, bloating, increased abdominal girth, urinary urgency or frequency, changes in bowel habits, and unexplained weight loss.*

LDH Isoenzymes

Lactate Dehydrogenase (LDH) is an enzyme found in almost all body tissues, with five main isoenzymes designated as LDH-1 through LDH-5.

The normal reference range for LDH levels varies depending on the laboratory and the method used for testing. LDH levels are typically higher in children and decrease with age.

Elevated levels

Tissue Injury and Disease

LDH is released into the bloodstream when cells are damaged, and its levels can increase in various pathological conditions.

Interpretation of LDH Isoenzymes

LDH isoenzyme analysis can help *differentiate the source of LDH elevation*, as each isoenzyme is predominantly found in specific tissues:

- LDH-1: Predominantly found in the *heart muscle and red blood cells.*
- LDH-2: Predominantly found in the *heart muscle.*
- LDH-3: Predominantly found in *lungs.*
- LDH-4: Predominantly found in *kidneys, placenta, and pancreas.*
- LDH-5: Predominantly found in the *liver and skeletal muscle.*

Elevated LDH Isoenzyme Levels

Elevated LDH-1 and LDH-2

Cardiac conditions such as *myocardial infarction (heart attack) or acute coronary syndrome* may result in increased LDH-1 and LDH-2 levels due to cardiac muscle damage.

Elevated LDH-3

Respiratory conditions such as *pneumonia, pulmonary embolism, or Acute Respiratory Distress Syndrome (ARDS)* may lead to increased LDH-3 levels due to lung tissue damage.

Elevated LDH-4 and LDH-5

Liver conditions such as hepatitis, liver cirrhosis, or liver metastases may result in increased LDH-4 and LDH-5 levels due to liver cell damage.

Non-specific LDH elevation

Some conditions, such as *hemolysis, malignancies, tissue necrosis, and autoimmune disorders,* can lead to non-specific elevation of LDH levels without a characteristic isoenzyme pattern.

Interpretation of LDH isoenzymes results requires consideration of clinical signs and symptoms, along with the specific pattern of LDH elevation.

Different isoenzyme patterns may provide clues to the underlying tissue damage or disease process, helping clinicians identify potential causes and guide further diagnostic evaluation and management.

Mammary-Specific Antigen

Mammary-Specific Antigen (MSA) is a protein found in the mammary glands, in breast tissue.

Normal reference range

MSA levels vary depending on the laboratory and the method used for testing.

Generally, MSA levels are low or undetectable in healthy individuals.

Elevated MSA levels

Breast Cancer Screening

MSA levels may be evaluated as part of breast cancer screening in women with suspected breast abnormalities or at high risk for breast cancer.

Clinical signs and symptoms associated with breast cancer may include *palpable breast lumps, changes in breast size or shape, skin dimpling or puckering, nipple discharge, nipple inversion, or axillary lymphadenopathy.*

Interpretation of MSA results requires consideration of the patient's clinical signs and symptoms, along with other diagnostic findings. Elevated MSA levels may raise suspicion for breast cancer, but further diagnostic evaluation is necessary to confirm the diagnosis and guide appropriate management.

PSA (Total and Free)

Interpreting the analytical results of PSA (prostate-specific antigen) in the context of clinical signs and symptoms involves considering various factors, including the patient's age, medical history, symptoms, and risk factors for prostate cancer.

Normal PSA Levels

Generally, a PSA level below 4.0 nanograms per millilitre (ng/mL) 1 for most men.

Elevated PSA Levels

Benign Prostatic Hyperplasia (BPH)

PSA levels can be elevated in men with BPH, a non-cancerous enlargement of the prostate gland.

Prostatitis

Inflammation or infection of the prostate gland can cause PSA levels to rise temporarily.

Prostate Cancer

Elevated PSA levels may be indicative of prostate cancer, especially if PSA levels continue to rise over time or if other risk factors are present.

PSA testing is a valuable tool for prostate cancer screening, but it is not diagnostic on its own.

Abnormal PSA levels should be interpreted in conjunction with other clinical findings and may require further evaluation to confirm a diagnosis and determine appropriate management.

PTHRP

Parathyroid Hormone-related Protein (PTHrP) is a protein that plays a role in calcium regulation and bone metabolism.

Normal reference range

PTHrP levels vary depending on the laboratory and the method used for testing.

In healthy individuals, PTHrP levels are typically low or undetectable.

Elevated PTHrP

Hypercalcemia

Elevated PTHrP levels are commonly associated with *hypercalcemia* (high levels of calcium in the blood).

Clinical signs and symptoms of hypercalcemia may include *fatigue, weakness, nausea, vomiting, constipation, increased thirst, frequent urination, abdominal pain, muscle weakness, confusion, and bone pain.*

Underlying Conditions

Elevated PTHrP levels may be seen in various medical conditions, including *malignancies (such as certain types of cancer, particularly solid tumors), hyperparathyroidism, Hypercalcemia of Malignancy (HCM), and certain non-cancerous conditions.*

Interpretation of PTHrP results involves consideration of the patient's clinical signs and symptoms, serum calcium levels, and underlying medical conditions. Elevated PTHrP levels are commonly associated with hypercalcemia of malignancy and may indicate the presence of an underlying malignancy, particularly in the context of clinical suspicion and other diagnostic findings.

PROTEIN ELECTROPHORESIS

Protein electrophoresis is a laboratory technique used to separate proteins based on their electrical charge and size. Interpretation of protein electrophoresis results should be considered in the context of the patient's clinical history, physical examination findings, other laboratory test results, additional diagnostic tests, such as serum protein immunofixation electrophoresis, serum free light chain assay, bone marrow biopsy, and imaging studies.

Normal protein fractions

Protein electrophoresis typically separates proteins into distinct fractions, including *albumin, alpha-1 globulins, alpha-2 globulins, beta globulins, and gamma globulins.*

Albumin is the most abundant protein fraction in the blood.

Other fractions primarily consist of various globulins, including immunoglobulins (antibodies).

Elevated levels

Clinical manifestations associated with *abnormal protein electrophoresis results* depend on the specific pattern of protein fractions and underlying conditions.

General signs and symptoms of abnormal protein levels may include *unexplained weight loss, fatigue, weakness, edema, easy bruising or bleeding, recurrent infections, bone pain, and organ dysfunction.*

Interpretation of Results

Hypoalbuminemia

Decreased albumin levels may indicate *liver disease, malnutrition, protein-losing enteropathy, nephrotic syndrome, or inflammation.*

Hyperglobulinemia

Elevated globulin levels, particularly the gamma globulin fraction, may indicate *chronic inflammation, autoimmune diseases, infectious diseases (e.g.*, chronic viral infections), hematological malignancies (*e.g.*, multiple myeloma, lymphoma), or chronic liver disease.

Monoclonal Bands

Presence of monoclonal bands, characterized by discrete spikes or peaks in the gamma globulin region, may indicate the presence of *monoclonal gammopathies* such as *multiple myeloma or Monoclonal Gammopathy of Undetermined Significance (MGUS).*

Polyclonal Bands

Diffuse elevation of gamma globulins with a polyclonal pattern may be seen in conditions such as *chronic infections, autoimmune diseases, liver diseases, or inflammatory disorders.*

Interpretation of protein electrophoresis results involves consideration of the specific pattern of protein fractions, presence of monoclonal or polyclonal bands, and clinical context to identify underlying diseases or conditions affecting protein metabolism and homeostasis. Prompt diagnosis and appropriate management are essential to address the underlying cause and optimize patient outcomes.

Thyroglobulin, Anti-Thyroglobulin Antibodies

Interpretation of Thyroglobulin (Tg) and anti-Thyroglobulin Antibodies (TgAb) results in the context of clinical signs and symptoms involves several considerations.

Normal reference ranges

Thyroglobulin (Tg)

Thyroglobulin is measured in the blood to assess thyroid function and monitor thyroid cancer recurrence.

Normal range for Tg varies depending on factors such as age, gender, and assay method.

Elevated Thyroglobulin (Tg)

Elevated Tg levels may indicate thyroid cancer recurrence or residual thyroid tissue, particularly *in patients who have undergone thyroidectomy (surgical removal of the thyroid gland) for thyroid cancer.*

Anti-Thyroglobulin Antibodies (TgAb)

TgAb are antibodies produced by the immune system that target thyroglobulin.

Elevated levels of TgAb may indicate *autoimmune thyroid disease, such as Hashimoto's thyroiditis*, or may interfere with Tg measurements.

Elevated TgAb levels *may interfere with Tg measurements*, leading to falsely elevated or inaccurate Tg results.

Concurrent Elevated Tg and TgAb

In some cases, elevated Tg levels in the presence of elevated TgAb may indicate ongoing thyroid tissue production or release, rather than thyroid cancer recurrence.

Additional diagnostic tests and clinical evaluation are necessary to confirm the underlying cause.

Interpretation of Tg and TgAb results should be considered in the context of *clinical history, thyroid cancer treatment history, thyroid function tests (e.g.*, TSH, free T4), imaging studies (*e.g.*, ultrasound, radioactive iodine scan), and other laboratory findings.

I*nterpretation of Thyroglobulin and anti-thyroglobulin antibodies results* requires consideration of clinical signs and symptoms, thyroid cancer treatment history, and other laboratory and imaging findings.

Elevated Tg levels may indicate *thyroid cancer recurrence*, while *elevated TgAb levels* may *interfere with Tg measurements* and require additional diagnostic evaluation.

Other Emerging Biomarkers

Interpreting the analytical results of emerging biomarkers in the context of clinical signs and symptoms involves several steps. Emerging biomarkers may include *novel proteins, genetic markers, metabolites, or other molecules* that show promise in predicting or diagnosing various conditions.

Emerging biomarkers of cancer are continually being discovered and studied to improve early detection, prognosis, treatment selection, and monitoring of cancer patients.

Interpretation of Results

The biomarker results were analyzed in comparison with established reference ranges or thresholds, if available.

Elevated or abnormal levels of emerging biomarkers may indicate underlying disease processes or physiological disturbances.

However, it is important to remember that the interpretation of emerging biomarkers may be more challenging due to limited data or validation compared to traditional biomarkers.

Clinical context

Consider the clinical context in which the biomarker testing is performed. This includes the patient's medical history, risk factors, concurrent medications, and other laboratory or imaging findings.

Integration of biomarker results into the broader clinical picture can provide valuable insights into disease diagnosis, prognosis, or treatment response.

Consider factors such as sensitivity, specificity, positive and negative predictive values, and potential limitations or confounders that may affect the accuracy and reliability of the biomarker measurements.

Consult

Where the interpretation of emerging biomarkers is uncertain or complex, *consultation with specialists*, such as *clinical chemists, pathologists, or other relevant healthcare professionals*, may be beneficial.

Collaborative decision-making based on multidisciplinary expertise can help ensure optimal patient care.

Interpreting the analytical results of emerging biomarkers requires a comprehensive approach that considers the specific biomarkers being measured, clinical signs and symptoms, available reference ranges or thresholds, clinical context, research evidence, and collaboration among healthcare providers (Table **11**).

Table 11. Interpretation of emerging biomarkers.

Biomarkers	*Application*
Circulating tumor DNA (ctDNA)	Provides information about tumor mutations, genetic alterations, and treatment response. ctDNA has the potential to be used for early cancer detection, monitoring of minimal residual disease, and assessment of treatment response and resistance.
Exosomes	Detection of exosomes in bodily fluids, such as blood, urine, or saliva, indicates that cells, including cancer cells, release these small extracellular vesicles into circulation. The presence of exosomes may suggest ongoing cellular communication, signaling, or biomolecular exchange within the body.
Tumor-derived Exosomes	Exosomes released by tumor cells, known as tumor-derived exosomes, carry a cargo of proteins, nucleic acids (including DNA, RNA, and microRNAs), lipids, and other biomolecules reflective of the parent tumor's molecular characteristics. Tumor-derived exosomes have diagnostic potential and act as therapeutic targets and may contribute to cancer progression, metastasis, immune evasion, and drug resistance.
Liquid biopsy markers	Involve analysis of Circulating Tumor Cells (CTCs), cell-free DNA (cfDNA), exosomes, and other biomarkers in blood or other bodily fluids. Less invasive than traditional tissue biopsies and can provide real-time information about tumor heterogeneity, treatment response, and disease progression.
Circulating Tumor Cells (CTCs)	Their presence suggests the presence of cancer cells that have detached from the primary tumor or metastatic lesions. The presence of CTCs may indicate tumor dissemination, metastatic potential, and disease progression.

(Table 11) cont.....

Biomarkers	*Application*
Cell-Free DNA (cfDNA) Analysis	Provide information about tumor-derived genetic mutations, copy number alterations, and chromosomal rearrangements. Identification of tumor-specific mutations or genetic aberrations in cfDNA may aid in cancer diagnosis, molecular profiling, and treatment selection.
Extracellular Vesicles (Exosomes)	Extracellular vesicles, including exosomes, contain bioactive molecules derived from tumor cells, such as proteins, nucleic acids, and lipids. Analysis of exosomal content may reveal biomarkers indicative of tumor subtype, stage, and treatment response.
MicroRNAs (miRNAs) and Non-coding RNAs	Dysregulated expression of specific miRNAs may serve as diagnostic or prognostic biomarkers for various cancer types.
Metabolic biomarkers	Metabolic biomarkers, such as *glucose, insulin, lipid profile (including cholesterol and triglycerides), and markers of inflammation (such as C-reactive protein),* provide insights into metabolic health and dysregulation. Abnormal levels of these biomarkers may indicate underlying metabolic disorders, such as *diabetes, dyslipidemia, insulin resistance, or metabolic syndrome.* Explored for their diagnostic, prognostic, and therapeutic implications in cancer.

This approach helps to ensure accurate interpretation and appropriate clinical management based on the emerging biomarker data [7, 8].

EMERGING BIOMARKERS IN CANCER RESEARCH

Summary

Interpreting the analytical results of liquid biopsy markers involves assessing the presence, quantity, and molecular characteristics of tumor-derived components in bodily fluids. Integration of liquid biopsy analysis with clinical signs and symptoms, imaging studies, and other diagnostic tests provides valuable insights into cancer diagnosis, prognosis, treatment response, and personalized therapeutic approaches.

Exercises

- What are oncogenes, and how do they contribute to cancer development?
- How do tumor suppressor genes function in preventing cancer? Give an example.
- Discuss the role of liquid biopsies in the early detection and monitoring of cancer. How do they compare with traditional tissue biopsies?
- How does understanding cancer metabolism contribute to the development of new cancer therapies? Provide examples of metabolic pathways targeted in cancer treatment.

- What is the primary function of the BRCA1 and BRCA2 genes?
- Which molecular technique is commonly used to detect gene mutations associated with cancer?

CONCLUSION

Overall, emerging biomarkers of cancer hold promise for improving cancer diagnosis, prognosis, and treatment by providing personalized and targeted approaches to cancer management. However, further research and validation are needed to establish the clinical utility and reliability of these biomarkers in routine clinical practice.

REFERENCES

[1] Fundamentals of Clinical Chemistry and Molecular Diagnostics, 9th Edition. Elsevier. **2026**.

[2] Kaplan, L.; Pesce, A. *Clinical Chemistry*; Mosby/Elsevier: St. Louis, Mo. 5th edition, **2010**.

[3] Murray, R.; Bender, D.; Botham, K.; Kennelly, P.; Rodwell, V.; Weil, P. Harpers Illustrated Biochemistry 33rd Edition. Lange; **2026**.

[4] Rifai, N. Fundamentals of Clinical Chemistry and Molecular Diagnostics, 8th Edition. Saunders. **2018**.

[5] Voet, D.; Voet, J.; Pratt, C. *Fundamentals Of Biochemistry*; 6th edition; **2025**.

[6] Hanahan, D.; Weinberg, R.A. Hallmarks of cancer: the next generation. *Cell,* **2011**, *144*(5), 646-674. [http://dx.doi.org/10.1016/j.cell.2011.02.013] [PMID: 21376230]

[7] Zhou, Y.; Tao, L.; Qiu, J. Tumor biomarkers for diagnosis, prognosis and targeted therapy *Sig Transduct Target Ther,* **2024**, *9*, 132. [http://dx.doi.org/10.1038/s41392-024-01823-2]

[8] Shuel, S.L. Targeted cancer therapies: Clinical pearls for primary care. *Can Fam Physician,* **2022**, *68*(7), 515-518. [http://dx.doi.org/10.46747/cfp.6807515]

CHAPTER 3

Cardiovascular Disorders and Hypertension Diseases

Abstract: Cardiovascular disorders like coronary artery disease, myocardial infarction, heart failure, and hypertension are major causes of illness and death globally. Laboratory testing is crucial for diagnosis and management. Key biomarkers such as troponins, BNP, lipids, and electrolytes play a vital role in clinical decision-making. Hypertension, a major risk factor for cardiovascular disease, requires careful monitoring through blood pressure measurements and specific laboratory tests. This module covers the biochemical basis of these diseases, laboratory diagnostics, and clinical approaches to patient management.

LEARNING OBJECTIVES

1. Describe the causes and manifestations of cardiovascular disease and the assessment of individual risk for cardiovascular disorders.
2. Describe the laboratory investigations that are important in the detection, diagnosis, and management.
3. Select and interpret the results of the laboratory tests and recognize the analytical methods available.

KEY CONCEPTS

CAUSES AND MANIFESTATIONS OF CARDIOVASCULAR DISEASE AND THE ASSESSMENT OF INDIVIDUAL RISK FOR CARDIOVASCULAR DISORDERS.

Acute Coronary Syndrome

Acute Coronary Syndrome (ACS) [1]

Def.: It refers to a range of conditions associated with sudden, reduced blood flow to the heart. *It includes conditions such as unstable angina, non-ST-segment Elevation Myocardial Infarction (NSTEMI), and ST-segment Elevation Myocardial Infarction (STEMI).*

Simmi Kharb

Biochemical Basis

Involves complex interactions between lipids, inflammatory processes, and hemodynamic factors (Table **1**).

Table 1. Biochemical basis of causes and clinical signs and symptoms of ACS.

Aspect (cause)	*Biochemical Basis*	*Causes*	*Clinical Signs and Symptoms*	*Biochemical Basis of Symptoms*
Atherosclerosis	Elevated LDL, Decreased HDL, Increased triglycerides	Chronic buildup of plaque in the coronary arteries	-	-
Plaque Rupture/Erosion	*Inflammatory mediators: e.g.,* CRP interleukins *Proteolytic enzymes: e.g.,* MMPs	Disruption of atherosclerotic plaque	-	-
Thrombus Formation	*Platelet activation*: Thromboxane A2 ADP Fibrinogen Coagulation cascade: Thrombin Fibrin	Formation of a blood clot obstructing blood flow in the coronary artery	-	-
Coronary Artery Spasm	*Imbalances in endothelial function:* Increased endothelin-1, Decreased NO	Sudden, temporary narrowing of a coronary artery	-	-
Chest Pain (Angina Pectoris)	Ischemia-induced release of adenosine and lactic acid	-	Sudden, severe, crushing chest pain, radiating to the neck, jaw, shoulder, or arm	Release of pain-inducing substances due to ischemia
Shortness of Breath (Dyspnea)	Myocardial ischemia and center ventricular dysfunction	-	Difficulty breathing	Pulmonary congestion and reduced oxygen delivery
Diaphoresis (Sweating)	Activation of the sympathetic nervous system	-	Excessive sweating	Increased sweat gland activity due to stress and pain
Nausea and Vomiting	Vagal nerve stimulation	-	Feeling of sickness with or without vomiting	Vagal response to myocardial ischemia and pain

(Table 1) cont.....

Aspect (cause)	*Biochemical Basis*	*Causes*	*Clinical Signs and Symptoms*	*Biochemical Basis of Symptoms*
Palpitations	Arrhythmias due to altered ion channel function and disrupted electrical conduction	-	Rapid, strong, or irregular heartbeat	Ischemia-induced electrical disturbances
Fatigue and Weakness	Reduced cardiac output and poor tissue perfusion	-	Unexplained tiredness or weakness	Decreased oxygen and nutrient delivery to tissues
Syncope (Fainting)	Severe reduction in cardiac output leading to decreased cerebral perfusion	-	Temporary loss of consciousness	Insufficient blood flow to the brain

Causes of ACS

1. Atherosclerosis
2. Plaque Rupture or Erosion
3. Thrombus Formation
4. Coronary Artery Spasm

Clinical Signs and Symptoms of Acute Coronary Syndrome

1. Chest Pain (Angina Pectoris): Sudden, severe, and crushing chest pain radiating to the neck, jaw, shoulder, or arm.
2. Shortness of Breath (Dyspnea): Difficulty breathing, which may accompany or follow chest pain.
3. Diaphoresis (Sweating): Excessive sweating.
4. Nausea and Vomiting: Feeling of sickness with or without vomiting.
5. Palpitations: Feeling of rapid, strong, or irregular heartbeat.
6. Fatigue and Weakness: Unexplained tiredness or weakness.
7. Syncope (Fainting): Temporary loss of consciousness.

KEY CONCEPTS: BIOCHEMICAL BASIS OF CAUSES AND CLINICAL SIGNS AND SYMPTOMS OF ACS

Key Concepts: Biochemical Changes of Disease or Disorder

The biochemical changes in ACS are a complex interplay of myocardial injury, inflammation, lipid metabolism, coagulation, oxidative stress, and endothelial dysfunction (Table **2**).

Table 2. Biochemical Changes in Acute Coronary Syndrome (ACS)[1 - 3].

Biochemical Change	*Description*	*Clinical Significance*
Cardiac Biomarkers		
Troponins cTnI and cTnT	Proteins released from damaged cardiac myocytes Highly specific to cardiac tissue.	Elevated levels indicate myocardial injury; Critical for the diagnosis of Myocardial Infarction (MI). (Fig. 3.1)
Creatine Kinase-MB (CK-MB)	Isoenzyme of creatine kinase found in cardiac muscle.	Elevated levels indicate myocardial damage Less specific than troponins.
Myoglobin	Oxygen-binding protein in muscle; released rapidly after muscle injury.	An early marker of muscle injury, but not specific to cardiac tissue.
Inflammatory Markers		
C-Reactive Protein (CRP)	An acute-phase protein produced by the liver in response to inflammation.	Elevated levels indicate systemic inflammation and are associated with an increased risk of plaque rupture.
Interleukins *e.g.*, IL-6	Cytokines involved in the inflammatory response.	Elevated levels reflect inflammatory activity within atherosclerotic plaques.
Tumor Necrosis Factor-alpha (TNF-α)	Pro-inflammatory cytokine.	Contributes to systemic and local inflammation in atherosclerotic plaques.
Lipid metabolism markers		
Low-Density Lipoprotein (LDL)	"Bad" cholesterol that contributes to plaque formation.	High levels are a major risk factor for atherosclerosis and subsequent ACS.
High-Density Lipoprotein (HDL)	"Good" cholesterol that helps remove cholesterol from arteries.	Low levels are associated with a higher risk of atherosclerosis and ACS.
Triglycerides	Type of fat found in the blood.	Elevated levels contribute to atherosclerosis.
Coagulation Factors		
Fibrinogen	Plasma protein involved in blood clotting.	Elevated levels increase the risk of coronary artery thrombosis.
D-dimer	Fibrin degradation products present after a blood clot is dissolved.	Elevated levels indicate recent clot formation and breakdown.
Oxidative stress markers		
Malondialdehyde (MDA)	Product of lipid peroxidation.	Elevated levels indicate oxidative stress and lipid damage within the arterial walls.
Superoxide Dismutase (SOD)	An enzyme that protects against oxidative damage by converting superoxide radicals to oxygen and hydrogen peroxide.	Decreased activity reflects impaired antioxidant defense.

(Table 2) cont.....

Biochemical Change	*Description*	*Clinical Significance*
Glutathione Peroxidase (GPx)	An enzyme that reduces hydrogen peroxide to water, protecting cells from oxidative damage.	Decreased activity indicates compromised antioxidant defense.
Electrolyte imbalance		
Potassium (K^+)	Essential for normal cardiac electrical activity.	Abnormal levels can lead to arrhythmias.
Magnesium (Mg^{2+})	Important for normal cardiac function and electrical stability.	Low levels can predispose to arrhythmias and other cardiac complications.
Endothelial dysfunction markers		
Nitric Oxide (NO)	A vasodilator produced by endothelial cells.	Reduced production leads to vasoconstriction and promotes thrombosis.
Endothelin-1	Potent vasoconstrictor peptide produced by endothelial cells.	Elevated levels contribute to vasospasm and atherosclerosis progression.
Metabolic markers		
Lactate	Produced during anaerobic metabolism.	Elevated levels indicate tissue hypoxia and anaerobic metabolism in ischemic myocardium.
Glucose	Elevated levels due to stress response and insulin resistance in ACS.	Hyperglycemia is associated with worse outcomes in ACS patients.

LABORATORY INVESTIGATIONS FOR DETECTION, DIAGNOSIS, AND MANAGEMENT

Monitoring these markers helps in the diagnosis, risk stratification, and management of patients with ACS.

Cardiac Risk Assessment[1-4]

Cardiac risk assessment involves evaluating various biochemical markers and using specific tools to predict the likelihood of cardiovascular events such as myocardial infarction (Tables **3**, **4**).

Table 3. Key biochemical markers and assessment tools used in cardiac risk assessment: Biochemical markers.

Marker	*Description*	*Clinical significance*
Lipid profile		
- Total Cholesterol	A measure of all cholesterol in the blood.	High levels are associated with increased risk of Coronary Artery Disease (CAD).
- LDL Cholesterol (LDL-C)	"Bad" cholesterol that contributes to plaque buildup in arteries.	Elevated levels are a major risk factor for atherosclerosis and cardiovascular events.
- HDL Cholesterol (HDL-C)	"Good" cholesterol that helps remove cholesterol from arteries.	Low levels are associated with a higher risk of cardiovascular disease.
- Triglycerides	Type of fat found in the blood.	Elevated levels are associated with increased risk of CAD and pancreatitis.
Inflammatory Markers		
- C-Reactive Protein (CRP)	An acute-phase protein that indicates inflammation.	High-sensitivity CRP (hs-CRP) is used to assess cardiovascular risk; elevated levels suggest a higher risk.
- Interleukin-6 (IL-6)	Cytokine involved in the inflammatory response.	Elevated levels are associated with increased cardiovascular risk.
Glycemic Control		
- Fasting Blood Glucose	Measure of blood sugar levels after fasting.	Elevated levels indicate diabetes or prediabetes, which are risk factors for cardiovascular disease.
- Hemoglobin A1c (HbA1c)	Reflects average blood glucose levels over the past 2-3 months.	Elevated levels indicate poor glycemic control and increased cardiovascular risk.
Lipoprotein Subfractions		
- Apolipoprotein B (ApoB)	Major protein component of LDL and VLDL particles.	High levels are associated with increased risk of atherosclerosis.
- Lipoprotein(a) [Lp(a)]	LDL particle with an additional protein (apolipoprotein(a)).	Elevated levels are an independent risk factor for cardiovascular disease.
Coagulation Markers		
- Fibrinogen	A plasma protein involved in blood clotting.	Elevated levels are associated with increased risk of thrombosis and cardiovascular events.
- D-dimer	Fibrin degradation product present after clot breakdown.	Elevated levels indicate recent clot formation and breakdown.
Renal Function Markers		
- Creatinine	Waste product of muscle metabolism, indicating kidney function.	Elevated levels indicate impaired kidney function, which is a risk factor for cardiovascular disease.

(Table 3) cont.....

Marker	*Description*	*Clinical significance*
Lipid profile		
- Albuminuria	Presence of albumin in urine, indicating kidney damage.	Associated with increased cardiovascular risk.

Table 4. Assessment Tools [4].

Tool	*Description*	*Clinical Use*
Framingham Risk Score	Algorithm based on age, gender, total cholesterol, HDL cholesterol, smoking status, and blood pressure.	Estimates the 10-year risk of developing coronary heart disease.
Reynolds Risk Score	Includes traditional risk factors plus hs-CRP and family history of premature heart disease.	Provides an estimate of 10-year cardiovascular risk, particularly in women.
Atherosclerotic Cardiovascular Disease (ASCVD) Risk Calculator	Uses pooled cohort equations to estimate the 10-year risk of heart attack and stroke.	Incorporates age, gender, race, total cholesterol, HDL cholesterol, systolic blood pressure, diabetes, and smoking status.
Coronary Artery Calcium (CAC) Score	Uses a CT scan to measure the amount of calcium in the walls of the coronary arteries.	Higher scores indicate a greater risk of coronary artery disease.
Carotid Intima-Media Thickness (CIMT)	Ultrasound measurement of the thickness of the carotid artery walls.	Increased thickness is associated with higher cardiovascular risk.
Ankle-Brachial Index (ABI)	Ratio of the blood pressure in the ankle to the blood pressure in the arm.	Lower ratios indicate peripheral artery disease and higher cardiovascular risk.
Genetic Risk Scores	Based on the presence of multiple genetic variants associated with cardiovascular disease.	Helps in assessing inherited risk, particularly in individuals with a family history of heart disease.
Stress Testing	Evaluates cardiovascular response to physical stress or pharmacologic agents.	Identifies ischemia, arrhythmias, and other cardiac abnormalities.
Echocardiography	Uses ultrasound waves to create images of the heart.	Assesses heart structure and function, including wall motion abnormalities indicative of ischemia.
Biomarker Panels	Comprehensive panels measuring multiple biomarkers (lipids, inflammatory markers, glycemic markers, *etc*).	Provides a broad overview of metabolic and inflammatory status to refine risk assessment.

Assessment Tools

The biochemical markers and assessment tools outlined above are essential for comprehensive cardiac risk assessment.

They help clinicians identify individuals at high risk of cardiovascular events, allowing for early intervention and management to prevent adverse outcomes.

Myocardial Infarct

Biochemical changes in myocardial infarction [1]

Cardiac Biomarkers:

- *Troponins* (cTnI and cTnT): Proteins released from damaged cardiac myocytes; elevated levels indicate myocardial injury.
- *Creatine Kinase-MB* (CK-MB): Isoenzyme of creatine kinase found in cardiac muscle; elevated levels indicate myocardial damage.
- *Myoglobin*: Oxygen-binding protein in muscle; released rapidly after muscle injury, but not specific to cardiac tissue.

Inflammatory Markers:

- *C-Reactive Protein (CRP*): Acute phase protein produced by the liver in response to inflammation; elevated levels indicate systemic inflammation.
- *Interleukins* (*e.g.*, IL-6): Cytokines involved in inflammatory response; elevated levels reflect inflammatory activity within atherosclerotic plaques.
- *Tumor Necrosis Factor-alpha (TNF-α):* Pro-inflammatory cytokine; contributes to systemic and local inflammation in atherosclerotic plaques.

Coagulation Markers:

- Fibrinogen: Plasma protein involved in blood clotting; elevated levels increase the risk of thrombosis in coronary arteries.
- *D-dimer:* Fibrin degradation product present after a blood clot is dissolved; elevated levels indicate recent clot formation and breakdown.

Lipid Metabolism Markers:

o *Low-Density Lipoprotein (LDL):* "Bad" cholesterol that contributes to plaque formation; high levels are a major risk factor for atherosclerosis.
o *High-Density Lipoprotein (HDL):* "Good" cholesterol that helps remove cholesterol from arteries; low levels are associated with a higher risk of atherosclerosis.

Oxidative Stress Markers:

o *Malondialdehyde (MDA):* Product of lipid peroxidation; elevated levels indicate oxidative stress and lipid damage within the arterial walls.
o *Superoxide Dismutase (SOD):* Enzyme that protects against oxidative damage; decreased activity reflects impaired antioxidant defense.
o *Glutathione Peroxidase (GPx):* Enzyme that reduces hydrogen peroxide to water, protecting cells from oxidative damage; decreased activity indicates compromised antioxidant defense.

Electrolyte Imbalance:

o *Potassium (K+):* Essential for normal cardiac electrical activity; abnormal levels can lead to arrhythmias.
o *Magnesium (Mg2+):* Important for normal cardiac function and electrical stability; low levels can predispose to arrhythmias.

Metabolic Markers:

o *Lactate:* Produced during anaerobic metabolism; elevated levels indicate tissue hypoxia and anaerobic metabolism in ischemic myocardium.
o *Glucose*: Elevated levels due to stress response and insulin resistance in myocardial infarction; hyperglycemia is associated with worse outcomes.

Myocardial infarction is primarily caused by atherosclerosis, plaque rupture, thrombus formation, and coronary artery spasm, leading to myocardial ischemia and infarction. The clinical signs and symptoms, such as chest pain, shortness of breath, and palpitations, are direct results of these biochemical and pathophysiological changes.

Biochemical markers like cardiac troponins, inflammatory markers, lipid profile, and coagulation factors play a crucial role in diagnosing and assessing the severity of myocardial infarction. Monitoring these biochemical changes is essential for timely intervention and management of the disease.

CONDITIONS

Relevant Laboratory Tests, Analytics Methods Available and Applied Aspects

Selection and interpretation of results of laboratory tests and recognize the available analytical methods.

Atherosclerosis

Relevant Laboratory Tests for Detection, Diagnosis, Management, and Monitoring are Mentioned in Table 5

Table 5. Relevant laboratory tests.

Laboratory Test	*Purpose*	*Biochemical Basis*	*Clinical Relevance*
Lipid Profile	Detection, Diagnosis	Measures levels of LDL, HDL, total cholesterol, and triglycerides.	Identifies dyslipidemia, a major risk factor for atherosclerosis. Provides targets for therapy.
Low-Density Lipoprotein (LDL)	Detection, Diagnosis, Monitoring	Elevated levels contribute to plaque formation in arteries.	High levels are associated with increased risk of atherosclerosis and cardiovascular disease.
High-Density Lipoprotein (HDL)	Detection, Diagnosis, Monitoring	Helps remove cholesterol from the arteries.	Low levels are associated with a higher risk of atherosclerosis.
Total Cholesterol	Detection, Diagnosis, Monitoring	Sum of LDL, HDL, and VLDL cholesterol.	Elevated levels indicate increased risk of atherosclerosis.
Triglycerides	Detection, Diagnosis, Monitoring	Elevated levels contribute to atherosclerosis and cardiovascular disease.	High levels are a risk factor for atherosclerosis and pancreatitis.
C-Reactive Protein (CRP)	Diagnosis, Monitoring	An acute-phase protein produced by the liver in response to inflammation.	Elevated levels indicate systemic inflammation and higher cardiovascular risk.
Homocysteine	Detection, Diagnosis	Elevated levels can damage the endothelium and promote plaque formation.	High levels are associated with increased risk of cardiovascular diseases.
Lipoprotein(a) [Lp(a)]	Detection, Diagnosis, Monitoring	A variant of LDL with an additional protein (apolipoprotein(a)); promotes atherogenesis.	High levels are an independent risk factor for cardiovascular diseases.

(Table 5) cont.....

Laboratory Test	***Purpose***	***Biochemical Basis***	***Clinical Relevance***
Apolipoprotein B (ApoB)	Detection, Diagnosis, Monitoring	Main protein component of LDL and VLDL; high levels reflect the number of atherogenic particles.	High levels indicate increased risk of atherosclerosis and cardiovascular disease.
Apolipoprotein A1 (ApoA1)	Detection, Diagnosis, Monitoring	Main protein component of HDL; low levels reflect poor reverse cholesterol transport.	Low levels are associated with increased risk of cardiovascular diseases.
Fibrinogen	Diagnosis, Monitoring	Plasma protein involved in blood clotting; high levels increase thrombotic risk.	Elevated levels are associated with increased risk of atherosclerosis and thrombosis.
D-dimer	Diagnosis, Monitoring	Fibrin degradation product; elevated levels indicate recent clot formation and breakdown.	High levels suggest a hypercoagulable state and risk of thrombotic events.
Oxidized LDL (oxLDL)	Detection, Diagnosis	LDL particles that have undergone oxidative modification; promotes inflammation and atherogenesis.	Elevated levels are associated with atherosclerosis progression and instability of plaques.
Blood Glucose (Fasting/Random)	Detection, Monitoring	Elevated levels indicate insulin resistance or diabetes mellitus, which contribute to atherosclerosis.	Diabetes and poor glycemic control are major risk factors for atherosclerosis.
Glycated Hemoglobin (HbA1c)	Monitoring	Reflects average blood glucose levels over the past 2-3 months.	Used to monitor long-term glycemic control in diabetic patients.
Inflammatory Cytokines (*e.g.*, IL-6, TNF-α)	Detection, Monitoring	Markers of systemic inflammation.	Elevated levels are associated with increased inflammation and risk of atherosclerosis.
High-Sensitivity CRP (hs-CRP)	Detection, Monitoring	A more sensitive measure of CRP levels.	Used to evaluate cardiovascular risk in apparently healthy individuals.
Serum Creatinine	Monitoring	Elevated levels indicate kidney dysfunction, which is a risk factor for cardiovascular disease.	Monitoring kidney function in patients with cardiovascular disease or risk factors.

Cardiac amyloid

Relevant Laboratory Tests for Detection, Diagnosis, Management, and Monitoring are presented in Table 6

Table 6. Laboratory tests for detection, diagnosis, management, and monitoring.

Laboratory Test	Purpose	Biochemical Basis	Clinical Relevance
Serum and Urine Protein Electrophoresis (SPEP and UPEP)	Detection, Diagnosis	Identifies abnormal monoclonal proteins (M-proteins) produced by plasma cells.	Detects light chain amyloidosis (AL amyloidosis).
Immunofixation Electrophoresis (IFE)	Diagnosis, Monitoring	Identifies and characterizes specific monoclonal proteins in serum and urine.	Confirms presence of monoclonal proteins.
Serum Free Light Chain Assay	Detection, Diagnosis, Monitoring	Measures levels of free kappa and lambda light chains.	Elevated levels indicate light chain amyloidosis.
Natriuretic Peptides (BNP, NT-proBNP)	Detection, Monitoring	Released by cardiac myocytes in response to ventricular stretching and pressure overload.	Elevated levels indicate cardiac dysfunction.
Cardiac Troponins (cTnI, cTnT)	Detection, Diagnosis, Monitoring	Released from damaged cardiac myocytes.	Elevated levels indicate myocardial injury.
Serum Amyloid A (SAA)	Monitoring	An acute-phase protein that increases in response to inflammation.	Elevated levels indicate systemic amyloidosis activity.
Genetic Testing	Diagnosis	Identifies mutations in genes associated with hereditary amyloidosis (*e.g.*, TTR gene for ATTR amyloidosis).	Confirms diagnosis of hereditary forms of amyloidosis.
Echocardiography	Diagnosis, Monitoring	Uses ultrasound to assess cardiac structure and function.	Detects characteristic features of cardiac amyloidosis, such as thickened ventricular walls.
Cardiac Magnetic Resonance Imaging (MRI)	Diagnosis, Monitoring	Provides detailed images of cardiac structure and tissue characteristics.	Detects amyloid deposits in the heart and assesses cardiac function.
Endomyocardial Biopsy	Diagnosis	Histological examination of cardiac tissue.	Confirms the presence of amyloid deposits in the myocardium.

(Table 6) cont.....

Laboratory Test	Purpose	Biochemical Basis	Clinical Relevance
Electrocardiogram (ECG)	Detection, Monitoring	Records the electrical activity of the heart.	Detects conduction abnormalities and low voltage QRS complex, typical of cardiac amyloidosis.
Serum and Urine Immunofixation	Detection, Diagnosis	Identifies abnormal proteins.	Used to confirm light chain amyloidosis.
24-hour Urine Protein	Monitoring	Measures total protein excreted in urine over 24 hours.	Elevated protein levels indicate renal involvement in amyloidosis.
Liver Function Tests (LFTs)	Monitoring	Measures levels of liver enzymes and proteins.	Assesses liver involvement in systemic amyloidosis.
Complete Blood Count (CBC)	Monitoring	Assesses overall health and detects a variety of disorders, including anemia.	Monitors for complications and side effects of treatment.

Congestive Heart Failure

Relevant Laboratory Tests for Detection, Diagnosis, Management, and Monitoring are presented in Table 7.

Table 7. Laboratory tests for detection, diagnosis, management, and monitoring [2, 3].

Laboratory Test	Purpose	Biochemical Basis	Clinical Relevance
Natriuretic Peptides (BNP, NT-proBNP)	Detection, Diagnosis, Monitoring	Released by cardiac myocytes in response to ventricular stretching and pressure overload.	Elevated levels indicate heart failure severity.
Cardiac Troponins (cTnI, cTnT)	Detection, Diagnosis, Monitoring	Released from damaged cardiac myocytes.	Elevated levels suggest myocardial injury or stress.
Serum Electrolytes (Sodium, Potassium, Chloride)	Monitoring	Imbalances can affect cardiac function and are common in heart failure due to medication effects or disease progression.	Monitors electrolyte status to manage heart failure treatment.
Blood Urea Nitrogen (BUN) and Creatinine	Monitoring	Evaluates kidney function, which can be impaired in heart failure.	Helps assess renal function and guide treatment decisions.
Liver Function Tests (LFTs)	Monitoring	Measures levels of liver enzymes and proteins.	Assesses liver congestion and dysfunction due to heart failure.

(Table 7) cont.....

Laboratory Test	Purpose	Biochemical Basis	Clinical Relevance
Complete Blood Count (CBC)	Monitoring	Assesses overall health and detects anemia or infection.	Monitors for anemia, which can worsen heart failure symptoms.
Thyroid Function Tests (TFTs)	Detection, Diagnosis	Evaluates thyroid hormone levels.	Thyroid dysfunction can contribute to heart failure.
Lipid Profile	Detection, Diagnosis, Monitoring	Measures levels of LDL, HDL, total cholesterol, and triglycerides.	Identifies dyslipidemia, a risk factor for cardiovascular diseases.
C-Reactive Protein (CRP)	Monitoring	An acute-phase protein produced by the liver in response to inflammation.	Elevated levels indicate systemic inflammation, which can exacerbate heart failure.
Echocardiography	Diagnosis, Monitoring	Uses ultrasound to assess cardiac structure and function.	Detects heart failure-related structural abnormalities such as center ventricular hypertrophy and reduced ejection fraction.
Electrocardiogram (ECG)	Detection, Monitoring	Records the electrical activity of the heart.	Identifies arrhythmias and other electrical abnormalities in heart failure.
Chest X-ray	Detection, Diagnosis	Visualizes heart size and pulmonary congestion.	Detects cardiomegaly and pulmonary edema.
Serum Ferritin and Transferrin Saturation	Monitoring	Assesses iron status, as both iron deficiency and overload can impact heart function.	Important for managing anemia and optimizing heart function in heart failure patients.
Urinalysis	Monitoring	Evaluates kidney function and detects proteinuria.	Monitors for renal complications associated with heart failure.
Arterial Blood Gases (ABGs)	Monitoring	Measures oxygen and carbon dioxide levels in the blood.	Assesses respiratory status in advanced heart failure.
Vitamin D Levels	Monitoring	Low levels are common in heart failure patients.	Vitamin D deficiency can worsen heart failure and affect overall health.
Glycated Hemoglobin (HbA1c)	Monitoring	Reflects average blood glucose levels over the past 2-3 months.	Monitors glycemic control in diabetic patients, a common comorbidity in heart failure.
Serum Albumin	Monitoring	Low levels indicate poor nutritional status or chronic disease.	Hypoalbuminemia is associated with worse outcomes in heart failure.

4. Hypertension[4]

Relevant Laboratory Tests for Detection, Diagnosis, Management, and Monitoring are presented in Table 8

Table 8. Laboratory tests for detection, diagnosis, management, and monitoring.

Laboratory Test	Purpose	Biochemical Basis	Clinical Relevance
Complete Blood Count (CBC)	Detection, Monitoring	Assesses overall health and detects anemia or infection.	Anemia can exacerbate hypertension and affect treatment.
Electrolyte Panel (Sodium, Potassium, Chloride)	Monitoring	Imbalances can contribute to hypertension or indicate kidney dysfunction.	Evaluates electrolyte balance and renal function.
Renal Function Tests (Creatinine, Blood Urea Nitrogen)	Monitoring	Evaluates kidney function, which is crucial for regulating blood pressure.	Detects renal impairment, a common cause of secondary hypertension.
Lipid Profile	Monitoring	Measures levels of cholesterol and triglycerides.	Dyslipidemia is a risk factor for cardiovascular diseases, including hypertension.
Glucose (Fasting Blood Sugar or Hemoglobin A1c)	Monitoring	Assesses blood glucose levels to detect diabetes, a risk factor for hypertension.	Diabetes is closely linked with hypertension and affects treatment decisions.
C-Reactive Protein (CRP)	Monitoring	Measures inflammation, which can contribute to hypertension.	Elevated levels indicate increased cardiovascular risk.
Thyroid Function Tests (TFTs)	Diagnosis, Monitoring	Evaluates thyroid hormone levels; thyroid dysfunction can affect blood pressure regulation.	Thyroid disorders can contribute to hypertension.
Urinalysis	Monitoring	Detects proteinuria or hematuria, which may indicate kidney damage associated with hypertension.	Monitors kidney function and assesses renal involvement in hypertension.
Serum Aldosterone and Renin Levels	Diagnosis, Monitoring	Assess the Renin-Angiotensi--Aldosterone System (RAAS) activity, which plays a role in blood pressure regulation.	Helps diagnose and manage secondary hypertension.

(Table 8) cont.....

Laboratory Test	Purpose	Biochemical Basis	Clinical Relevance
Serum Magnesium	Monitoring	Low levels of magnesium can contribute to hypertension.	Evaluates magnesium status, which may affect blood pressure regulation.
Serum Calcium	Monitoring	Abnormal calcium levels can affect blood vessel constriction and blood pressure.	Evaluates calcium status and its role in hypertension.
Urinary Albumin-t--Creatinine Ratio (UACR)	Monitoring	Detects early signs of kidney damage associated with hypertension.	Identifies renal involvement in hypertensive patients.
24-Hour Urine Sodium	Monitoring	Measures sodium excretion, which can influence blood pressure regulation.	Provides insight into sodium intake and its impact on hypertension.
Genetic Testing (for Monogenic Hypertension)	Diagnosis	Identifies genetic mutations associated with rare forms of hypertension.	Helps diagnose monogenic forms of hypertension.
Ambulatory Blood Pressure Monitoring (ABPM)	Monitoring	Records blood pressure at regular intervals over 24 hours, providing a more accurate assessment of hypertension.	Helps diagnose and monitor hypertension, especially in ambulatory settings.
Echocardiography	Diagnosis, Monitoring	Uses ultrasound to assess heart structure and function, detecting complications of long-standing hypertension.	Evaluates cardiac health and complications of hypertension.
Cardiac Biomarkers (BNP, Troponins)	Monitoring	Indicators of cardiac strain or damage, which can occur in hypertensive heart disease.	Detects and monitors cardiac involvement in hypertension.

Myocardial infarction [1]

Relevant Laboratory Tests for Detection, Diagnosis, Management, and Monitoring are presented in Table 9

Table 9. Laboratory tests for detection, diagnosis, management, and monitoring.

Laboratory Test	Purpose	Biochemical Basis	Clinical Relevance
Cardiac Troponins (cTnI, cTnT)	Detection, Diagnosis, Monitoring	Released from damaged cardiac myocytes due to myocardial injury.	Elevated levels indicate myocardial damage and help diagnose myocardial infarction.

(Table 9) cont.....

Laboratory Test	Purpose	Biochemical Basis	Clinical Relevance
Creatine Kinase (CK) and CK-MB Fraction	Detection, Diagnosis, Monitoring	CK-MB isoenzyme specifically found in cardiac tissue; released during myocardial injury.	Elevated levels indicate myocardial damage but are less specific than troponins.
Myoglobin	Detection, Diagnosis, Monitoring	Released early in myocardial injury, but lacks specificity for cardiac tissue.	Rapid rise and fall; used in conjunction with troponins for early detection.
Complete Blood Count (CBC)	Monitoring	Detects anemia or infection.	Anemia may worsen myocardial oxygen supply-demand balance; infection can complicate recovery.
Electrolyte Panel	Monitoring	Imbalances can affect cardiac function and complicate recovery.	Critical for maintaining proper cardiac function; imbalances can worsen myocardial injury.
Lipid Profile	Monitoring	Assesses cholesterol levels; dyslipidemia is a risk factor for cardiovascular diseases, including myocardial infarction.	High cholesterol levels contribute to atherosclerosis and increase the risk of myocardial infarction.
Blood Glucose	Monitoring	Evaluates blood sugar levels.	Hyperglycemia is associated with worse outcomes in myocardial infarction.
C-Reactive Protein (CRP)	Monitoring	Measures inflammation, which can exacerbate myocardial injury.	Elevated levels indicate increased risk of complications and adverse outcomes.
Brain Natriuretic Peptide (BNP)	Monitoring	Released by the heart in response to increased ventricular wall stress.	Elevated levels indicate cardiac dysfunction and may predict complications post-MI.
D-Dimer	Diagnosis, Monitoring	Fragment of fibrin formed during blood clot dissolution; elevated in thrombotic events, including myocardial infarction.	Used to rule out thrombotic complications, such as pulmonary embolism, post-MI.
Liver Function Tests (LFTs)	Monitoring	Assesses liver function; liver damage can occur due to cardiac-related hypoperfusion or medications.	Helps assess overall health and potential complications post-MI.
Renal Function Tests	Monitoring	Evaluates kidney function; renal impairment is common post-MI due to hypoperfusion or contrast dye use.	Detects renal dysfunction, which can complicate management and increase mortality risk post-MI.
Thyroid Function Tests	Monitoring	Thyroid dysfunction can impact cardiovascular health and recovery post-MI.	Identifies thyroid disorders that may worsen outcomes post-MI.

(Table 9) cont.....

Laboratory Test	Purpose	Biochemical Basis	Clinical Relevance
Urinalysis	Monitoring	Detects kidney dysfunction or signs of infection.	Screens for renal complications and urinary tract infections post-MI.
Arterial Blood Gas (ABG)	Monitoring	Assesses acid-base status and oxygenation; critical in hemodynamically unstable patients post-MI.	Guides management in patients with respiratory distress or shock post-MI.
Coagulation Profile	Monitoring	Evaluates clotting factors and risk of bleeding or thrombosis post-MI.	Guides anticoagulant therapy and monitors for bleeding or clotting complications.
Electrocardiogram (ECG)	Diagnosis, Monitoring	Records the electrical activity of the heart; specific ECG changes indicate myocardial infarction.	Essential for diagnosis and ongoing monitoring of myocardial infarction.
Echocardiography	Diagnosis, Monitoring	Uses ultrasound to assess cardiac structure and function; detects ventricular dysfunction post-MI.	Evaluates cardiac function and identifies complications such as heart failure post-MI.
Imaging (MRI/CT)	Diagnosis, Monitoring	Provides detailed images of cardiac anatomy and function; identifies infarct size and complications post-MI.	Assists in diagnosis, prognosis, and guiding treatment decisions post-MI.
Genetic Testing	Diagnosis, Management	Identifies genetic mutations associated with familial hypercholesterolemia or rare cardiomyopathies that predispose to MI.	Helps identify underlying genetic predispositions and guide management strategies.

Stable angina

Relevant Laboratory Tests for Detection, Diagnosis, Management, and Monitoring are presented in Table 10

Table 10. Laboratory tests for detection, diagnosis, management, and monitoring.

Laboratory Test	Purpose	Biochemical Basis	Clinical Relevance
Lipid Profile	Detection, Diagnosis, Monitoring	Measures levels of cholesterol and triglycerides.	Dyslipidemia is a risk factor for atherosclerosis, which can lead to stable angina.
High-Sensitivity C-Reactive Protein (hs-CRP)	Detection, Monitoring	Measures systemic inflammation, which can contribute to atherosclerosis and plaque instability.	Elevated levels may indicate increased cardiovascular risk and the presence of underlying inflammation.

(Table 10) cont.....

Laboratory Test	Purpose	Biochemical Basis	Clinical Relevance
Complete Blood Count (CBC)	Monitoring	Detects anemia or infection.	Anemia may worsen myocardial oxygen supply-demand balance; infection can complicate recovery.
Electrolyte Panel	Monitoring	Imbalances can affect cardiac function and complicate recovery.	Critical for maintaining proper cardiac function; imbalances can worsen myocardial oxygenation.
Blood Glucose	Monitoring	Evaluates blood sugar levels.	Diabetes is a risk factor for atherosclerosis and can exacerbate stable angina symptoms.
Thyroid Function Tests	Monitoring	Thyroid dysfunction can impact cardiovascular health.	Identifies thyroid disorders that may contribute to stable angina symptoms.
Liver Function Tests (LFTs)	Monitoring	Assesses liver function; liver damage can affect medication metabolism and cardiovascular health.	Helps assess overall health and potential complications related to stable angina treatment.
Renal Function Tests	Monitoring	Evaluates kidney function; renal impairment can affect medication clearance and cardiovascular health.	Detects renal dysfunction, which may complicate management and affect prognosis in stable angina.
Urinalysis	Monitoring	Detects kidney dysfunction or signs of infection.	Screens for renal complications and urinary tract infections, which may affect the stable angina prognosis.
Electrocardiogram (ECG)	Diagnosis, Monitoring	Records the electrical activity of the heart; may show ST-segment depression or T-wave inversion during anginal episodes.	Essential for diagnosing stable angina and assessing myocardial ischemia during stress testing.
Exercise Stress Test (Treadmill Test)	Diagnosis, Monitoring	Evaluates cardiovascular response to exercise; may reveal exercise-induced ischemia manifested as ECG changes or symptoms.	Confirms the presence of stable angina and assesses its severity and functional capacity.
Echocardiography	Diagnosis, Monitoring	Uses ultrasound to assess cardiac structure and function; evaluates wall motion abnormalities and ventricular function.	Assesses cardiac function and identifies complications such as heart failure in stable angina patients.
Imaging (MRI/CT)	Diagnosis, Monitoring	Provides detailed images of cardiac anatomy and function; may detect coronary artery stenosis or myocardial ischemia.	Assists in diagnosis, prognosis, and guiding treatment decisions in stable angina.

(Table 10) cont.....

Laboratory Test	Purpose	Biochemical Basis	Clinical Relevance
Coronary Angiography	Diagnosis, Management	An invasive procedure to visualize coronary arteries and assess for stenosis or blockages.	Confirms the presence and severity of coronary artery disease, guiding treatment decisions in stable angina.
Genetic Testing	Diagnosis, Management	Identifies genetic mutations associated with familial hypercholesterolemia or other hereditary cardiovascular conditions.	Helps identify underlying genetic predispositions and guide management strategies for stable angina.

Unstable angina

Relevant Laboratory Tests for Detection, Diagnosis, Management, and Monitoring are presented in Table 11

Table 11. Laboratory tests for detection, diagnosis, management, and monitoring.

Laboratory Test	Purpose	Biochemical Basis	Clinical Relevance
Cardiac Troponins (cTnI, cTnT)	Detection, Diagnosis, Monitoring	Released from damaged cardiac myocytes due to myocardial injury.	Elevated levels indicate myocardial damage and help diagnose unstable angina.
High-Sensitivity C-Reactive Protein (hs-CRP)	Detection, Monitoring	Measures systemic inflammation, which can contribute to plaque destabilization and rupture.	Elevated levels may indicate increased cardiovascular risk and the presence of underlying inflammation.
Lipid Profile	Monitoring	Measures levels of cholesterol and triglycerides.	Dyslipidemia is a risk factor for atherosclerosis and can contribute to unstable angina.
Complete Blood Count (CBC)	Monitoring	Detects anemia or infection.	Anemia may worsen myocardial oxygen supply-demand balance; infection can complicate recovery.
Electrolyte Panel	Monitoring	Imbalances can affect cardiac function and complicate recovery.	Critical for maintaining proper cardiac function; imbalances can worsen myocardial oxygenation.
Blood Glucose	Monitoring	Evaluates blood sugar levels.	Diabetes is a risk factor for atherosclerosis and can exacerbate unstable angina symptoms.
Thyroid Function Tests	Monitoring	Thyroid dysfunction can impact cardiovascular health.	Identifies thyroid disorders that may contribute to unstable angina symptoms.

(Table 11) cont.....

Laboratory Test	Purpose	Biochemical Basis	Clinical Relevance
Liver Function Tests (LFTs)	Monitoring	Assesses liver function; liver damage can affect medication metabolism and cardiovascular health.	Helps assess overall health and potential complications related to unstable angina treatment.
Renal Function Tests	Monitoring	Evaluates kidney function; renal impairment can affect medication clearance and cardiovascular health.	Detects renal dysfunction, which may complicate management and affect prognosis in unstable angina.
Urinalysis	Monitoring	Detects kidney dysfunction or signs of infection.	Screens for renal complications and urinary tract infections, which may affect unstable angina prognosis.
Electrocardiogram (ECG)	Diagnosis, Monitoring	Records the electrical activity of the heart; may show ST-segment changes indicative of ischemia during anginal episodes.	Essential for diagnosing unstable angina and assessing myocardial ischemia.
Exercise Stress Test (Treadmill Test)	Diagnosis, Monitoring	Evaluates cardiovascular response to exercise; may reveal exercise-induced ischemia manifested as ECG changes or symptoms.	Confirms the presence of unstable angina and assesses its severity and functional capacity.
Echocardiography	Diagnosis, Monitoring	Uses ultrasound to assess cardiac structure and function; evaluates wall motion abnormalities and ventricular function.	Assesses cardiac function and identifies complications such as heart failure in unstable angina patients.
Imaging (MRI/CT)	Diagnosis, Monitoring	Provides detailed images of cardiac anatomy and function; may detect coronary artery stenosis or myocardial ischemia.	Assists in diagnosis, prognosis, and guiding treatment decisions in unstable angina.
Coronary Angiography	Diagnosis, Management	An invasive procedure to visualize coronary arteries and assess for stenosis or blockages.	Confirms the presence and severity of coronary artery disease, guiding treatment decisions in unstable angina.
Genetic Testing	Diagnosis, Management	Identifies genetic mutations associated with familial hypercholesterolemia or other hereditary cardiovascular conditions.	Helps identify underlying genetic predispositions and guide management strategies for unstable angina.

RELEVANT LABORATORY TESTS, ANALYTICS METHODS AVAILABLE AND APPLIED ASPECTS

High-sensitivity CRP Table 12

Table 12. Laboratory tests for detection, diagnosis, management, and monitoring.

High-sensitivity CRP	Applied aspects
Interpretation of Analytical Results	High levels indicate inflammation and increased cardiovascular risk.
In Terms of Clinical Signs and Symptoms of Disease	Elevated levels may suggest underlying inflammatory conditions or increased risk of cardiovascular events.
Analytical Methods Available for the Laboratory Tests	Enzyme-Linked Immunosorbent Assay (ELISA) and immunoturbidimetric assays are commonly used methods.

Relevant laboratory tests, analytics methods available, and applied aspects

Natriuretic peptides Table 13

Table 13. Laboratory tests for detection, diagnosis, management, and monitoring.

Natriuretic Peptides	Applied aspects
Interpretation of Analytical Results	Elevated levels indicate cardiac strain or heart failure.
In Terms of Clinical Signs and Symptoms of Disease	Increased levels may suggest heart failure, cardiac dysfunction, or volume overload.
Analytical Methods Available for the Laboratory Tests	Immunoassays such as Enzyme-Linked Immunosorbent Assay (ELISA) and Chemiluminescent Immunoassays (CLIA) are commonly used.

Relevant laboratory tests, analytics methods available and applied aspects

Serum and urine electrophoresis and immunoglobulins Table 14

Table 14. Laboratory tests for detection, diagnosis, management, and monitoring.

Serum and Urine Electrophoresis and Immunoglobulins	Applied aspects
Interpretation of Analytical Results	Identifies abnormal protein patterns, such as monoclonal gammopathy or multiple myeloma.
In Terms of Clinical Signs and Symptoms of Disease	Abnormal patterns may indicate hematological malignancies or immune system disorders.

(Table 14) cont.....

Serum and Urine Electrophoresis and Immunoglobulins	Applied aspects
Analytical Methods Available for the Laboratory Tests	Techniques include agarose gel electrophoresis, capillary electrophoresis, and immunofixation electrophoresis.

Relevant laboratory tests, analytics methods available, and applied aspects

Troponin and high-sensitivity troponin (Table 15, Fig. 1)

Table 15. Laboratory tests for detection, diagnosis, management, and monitoring.

Troponin and High-Sensitivity Troponin	Applied aspects
Interpretation of Analytical Results	Elevated levels indicate myocardial injury, such as myocardial infarction.
In Terms of Clinical Signs and Symptoms of Disease	Increased levels are associated with chest pain, myocardial infarction, or other acute coronary syndromes.
Analytical Methods Available for the Laboratory Tests	Conventional troponin assays use Enzyme-Linked Immunosorbent Assays (ELISA), while high-sensitivity troponin assays utilize more sensitive detection methods like chemiluminescence or electrochemiluminescence immunoassays.

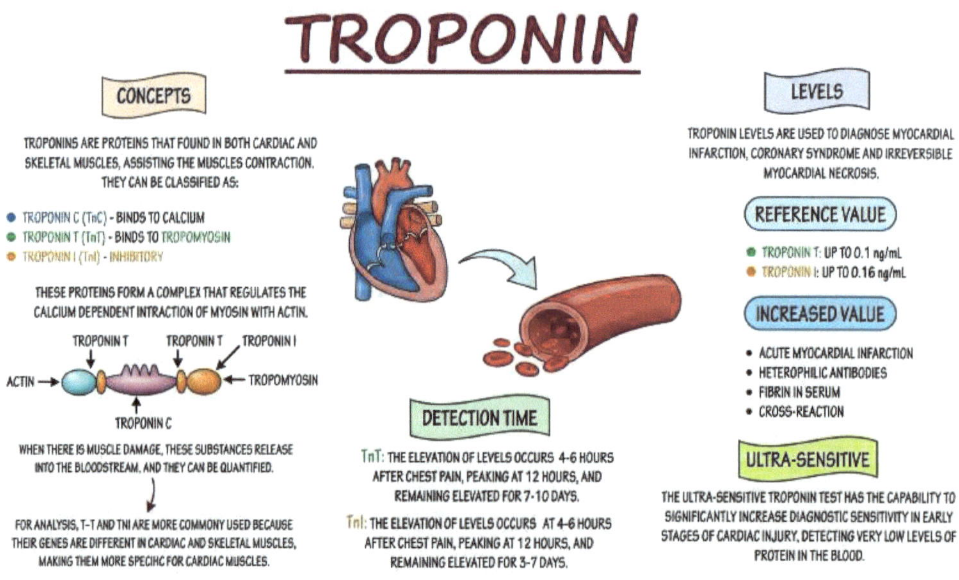

Fig. (1). Troponin.

Relevant laboratory tests, analytics methods available and applied aspects

Use and limitations of testing strategies/algorithms for cardiac markers Table 16 [4]

Table 16. Testing strategy/algorithm for cardiac markers.

Testing Strategy/Algorithm	Use	Limitations
Cardiac Markers for AMI and ACS	Identify myocardial injury or infarction.	Limited sensitivity in the early hours after symptom onset.
	Assist in diagnosing ACS and risk stratification.	Troponin elevation can occur in other conditions (*e.g.*, renal failure).
	Monitor response to treatment.	Creatine Kinase-MB (CK-MB) may lack specificity.
	Predict prognosis and guide management decisions.	Interpretation influenced by comorbidities and medications.
Lipid Screening for CHD	Assess cardiovascular risk.	Does not capture all risk factors (*e.g.*, family history, inflammation).
	Guide lipid-lowering therapy.	LDL cholesterol may not adequately reflect risk, especially in certain populations (*e.g.*, diabetics).
	Inform lifestyle modifications.	Misses non-lipid-related contributors to CHD risk (*e.g.*, smoking, hypertension).
	Monitor response to treatment.	Some lipid tests (*e.g.*, non-HDL cholesterol) may be more informative than others.
	Facilitate shared decision-making with patients.	Risk thresholds for treatment initiation are subject to debate.

Summary

Cardiovascular disorders are influenced by various factors, including genetic predisposition, lifestyle, and systemic diseases such as diabetes. The laboratory plays a crucial role in diagnosing conditions. Hypertension, a chronic condition characterized by elevated blood pressure, is diagnosed through blood pressure monitoring and biochemical tests that evaluate electrolyte balance, renal function, and the activity of the renin-angiotensin-aldosterone system.

Exercises

1. What are the primary laboratory markers for diagnosing myocardial infarction?
2. Discuss the role of troponins and the timing of their rise following an acute coronary event.

3. Explain how Brain Natriuretic Peptide (BNP) and NT-proBNP are used in diagnosing and managing heart failure.
4. How do these markers correlate with the severity of the disease?
5. In a patient with chest pain and elevated troponin levels, which additional laboratory and imaging tests should be done to confirm a diagnosis of myocardial infarction?
6. Describe the biochemical pathways involved in hypertension and how laboratory tests are used in diagnosis and management.
7. What is the role of lipid panels in cardiovascular risk assessment, and how can they be used to guide treatment?
8. In a hypertensive patient with low potassium and high aldosterone levels, what are the possible diagnoses, and how would you confirm them using laboratory tests?
9. Discuss how natriuretic peptides can be used to monitor the response to treatment in heart failure.

CONCLUSION

The management of cardiovascular disorders and hypertension is heavily reliant on accurate and timely biochemical diagnostics. Troponins, BNP, lipid panels, and electrolyte testing form the foundation of laboratory evaluations, helping to guide therapeutic decisions and assess patient outcomes. By integrating laboratory data with clinical findings, healthcare providers can tailor treatments to prevent complications such as stroke, myocardial infarction, and heart failure. Understanding the pathophysiology of hypertension and its impact on cardiovascular health is also crucial for long-term management and prevention of cardiovascular events.

REFERENCES

[1] Thygesen, K.; Alpert, J.S.; Jaffe, A.S.; Chaitman, B.R.; Bax, J.J.; Morrow, D.A.; White, H.D.; Mickley, H.; Crea, F.; Van de Werf, F.; Bucciarelli-Ducci, C.; Katus, H.A.; Pinto, F.J.; Antman, E.M.; Hamm, C.W.; De Caterina, R.; Januzzi, J.L., Jr; Apple, F.S.; Alonso Garcia, M.A.; Underwood, S.R.; Canty, J.M., Jr; Lyon, A.R.; Devereaux, P.J.; Zamorano, J.L.; Lindahl, B.; Weintraub, W.S.; Newby, L.K.; Virmani, R.; Vranckx, P.; Cutlip, D.; Gibbons, R.J.; Smith, S.C.; Atar, D.; Luepker, R.V.; Robertson, R.M.; Bonow, R.O.; Steg, P.G.; O'Gara, P.T.; Fox, K.A.A.; Hasdai, D.; Aboyans, V.; Achenbach, S.; Agewall, S.; Alexander, T.; Avezum, A.; Barbato, E.; Bassand, J-P.; Bates, E.; Bittl, J.A.; Breithardt, G.; Bueno, H.; Bugiardini, R.; Cohen, M.G.; Dangas, G.; de Lemos, J.A.; Delgado, V.; Filippatos, G.; Fry, E.; Granger, C.B.; Halvorsen, S.; Hlatky, M.A.; Ibanez, B.; James, S.; Kastrati, A.; Leclercq, C.; Mahaffey, K.W.; Mehta, L.; Müller, C.; Patrono, C.; Piepoli, M.F.; Piñeiro, D.; Roffi, M.; Rubboli, A.; Sharma, S.; Simpson, I.A.; Tendera, M.; Valgimigli, M.; van der Wal, A.C.; Windecker, S.; Chettibi, M.; Hayrapetyan, H.; Roithinger, F.X.; Aliyev, F.; Sujayeva, V.; Claeys, M.J.; Smajić, E.; Kala, P.; Iversen, K.K.; El Hefny, E.; Marandi, T.; Porela, P.; Antov, S.; Gilard, M.; Blankenberg, S.; Davlouros, P.; Gudnason, T.; Alcalai, R.; Colivicchi, F.; Elezi, S.; Baitova, G.; Zakke, I.; Gustiene, O.; Beissel, J.; Dingli, P.; Grosu, A.; Damman, P.; Juliebø, V.; Legutko, J.; Morais, J.; Tatu-Chitoiu, G.; Yakovlev, A.; Zavatta, M.; Nedeljkovic, M.; Radsel, P.; Sionis, A.; Jemberg, T.; Müller, C.; Abid, L.; Abaci, A.; Parkhomenko, A.; Corbett, S. ESC Scientific Document

Group. Fourth universal definition of myocardial infarction (2018). *Eur. Heart J.,* **2019**, *40*(3), 237-269.
[http://dx.doi.org/10.1093/eurheartj/ehy462] [PMID: 30165617]

[2] Hoevelmann, J.; Markwirth, P.; Tokcan, M.; Haring, B. What's new in heart failure? *Eur J Heart Fail,* **2025**, *27*(9), 1603-1605.
[http://dx.doi.org/10.1002/ejhf.70054]

[3] Berezin, A.E.; Berezin, A.A. Biomarkers in Heart Failure: From Research to Clinical Practice. *Ann. Lab. Med.,* **2023**, *43*(3), 225-236.
[http://dx.doi.org/10.3343/alm.2023.43.3.225] [PMID: 36544334]

[4] Whelton, P.K.; Carey, R.M.; Mancia, G.; Kreutz, R.; Bundy, J.D.; Williams, B. Harmonization of the American College of Cardiology/American Heart Association and European Society of Cardiology/European Society of Hypertension Blood Pressure/Hypertension Guidelines. *J. Am. Coll. Cardiol.,* **2022**, *80*(12), 1192-1201.
[http://dx.doi.org/10.1016/j.jacc.2022.07.005] [PMID: 35965201]

CHAPTER 4

Diabetes Mellitus

Abstract: Diabetes Mellitus (DM) is a chronic disorder characterized by high blood sugar levels due to issues with insulin secretion or action. Type 1 DM is autoimmune, while Type 2 DM is linked to insulin resistance and obesity. Diagnosis and monitoring involve tests. Advanced tests, such as C-peptide, help differentiate between Type 1 and Type 2 diabetes. Early detection and tight control of blood glucose levels are vital in preventing complications such as diabetic ketoacidosis, retinopathy, nephropathy, and cardiovascular diseases.

LEARNING OBJECTIVES

After going through this chapter, one should be able to:

- Describe the pathogenesis of diabetic states and the applied aspects of the study of diabetes mellitus.
- Describe the causes, clinical signs, and symptoms of applied aspects of diabetes mellitus.
- Describe the laboratory investigations that are important in the detection, diagnosis, and management of complications.
- Describe the specific laboratory investigations important to the study of diabetes mellitus.
- Recognize analytical methods available for their measurement.
- Select laboratory investigations and interpret the analytical results in the context of clinical signs and symptoms.

KEY CONCEPTS: PATHOGENESIS OF DIABETIC STATES AND REGULATION OF GLUCOSE HOMEOSTASIS

Pathogenesis of Diabetic States and the Applied Aspects of the Study of Diabetes Mellitus (Table 1) [1]

Table 1. Pathogenesis of diabetic states.

Aspect	Type 1 Diabetes Mellitus (T1DM)	Type 2 Diabetes Mellitus (T2DM)
Pathogenesis	- Autoimmune destruction of β-cells	- Insulin resistance in peripheral tissues (muscle, fat, liver)
	- Genetic susceptibility (*e.g.*, HLA-DR3, HLA-DR4)	- β-cell dysfunction over time
	- Environmental triggers (*e.g.*, viral infections, dietary factors, toxins)	- Strong association with obesity, sedentary lifestyle, and genetic predisposition
Insulin	- Secreted by pancreatic β-cells	- Secreted by pancreatic β-cells
Primary Function	- Lowers blood glucose levels	- Lowers blood glucose levels
Mechanisms	- Promotes glucose uptake in muscle and adipose tissue	- Promotes glucose uptake in muscle and adipose tissue
	- Stimulates glycogenesis (conversion of glucose to glycogen)	- Stimulates glycogenesis (conversion of glucose to glycogen)
	- Encourages lipogenesis (fatty acid synthesis) and inhibits lipolysis	- Encourages lipogenesis (fatty acid synthesis) and inhibits lipolysis
	- Enhances protein synthesis and inhibits protein degradation	- Enhances protein synthesis and inhibits protein degradation
Glucagon	- Secreted by pancreatic α-cells	- Secreted by pancreatic α-cells
Primary Function	- Raises blood glucose levels	- Raises blood glucose levels
Mechanisms	- Stimulates glycogenolysis (breakdown of glycogen to glucose)	- Stimulates glycogenolysis (breakdown of glycogen to glucose)
	- Promotes gluconeogenesis (synthesis of glucose from non-carbohydrate substrates)	- Promotes gluconeogenesis (synthesis of glucose from non-carbohydrate substrates)
Other Hormones Involved	- Epinephrine: Increases blood glucose by promoting glycogenolysis and gluconeogenesis	- Epinephrine: Increases blood glucose by promoting glycogenolysis and gluconeogenesis
	- Cortisol: Enhances gluconeogenesis and inhibits glucose uptake in tissues	- Cortisol: Enhances gluconeogenesis and inhibits glucose uptake in tissues
	- Growth Hormone: Reduces glucose uptake in tissues and increases lipolysis	- Growth Hormone: Reduces glucose uptake in tissues and increases lipolysis
Regulatory Pathways	- Fed State: High blood glucose levels stimulate insulin release	- Fed State: High blood glucose levels stimulate insulin release
	- Fasting State: Low blood glucose levels stimulate glucagon release	- Fasting State: Low blood glucose levels stimulate glucagon release
Dysregulation in Diabetes	- Lack of insulin leads to unopposed glucagon action	- Insulin resistance and β-cell dysfunction result in impaired glucose uptake

(Table 1) cont.....

Aspect	Type 1 Diabetes Mellitus (T1DM)	Type 2 Diabetes Mellitus (T2DM)
	- Causes hyperglycemia, ketogenesis, and metabolic acidosis	- Causes increased hepatic glucose production and chronic hyperglycemia
Clinical Implications	- Diagnosis: Based on fasting glucose levels, HbA1c, and glucose tolerance tests	- Diagnosis: Based on fasting glucose levels, HbA1c, and glucose tolerance tests
	- Management: Insulin therapy is essential	- Management: Lifestyle modification, oral hypoglycemic agents, insulin therapy
	- Complications: Leads to microvascular (retinopathy, nephropathy, neuropathy) and macrovascular (cardiovascular disease) complications	- Complications: Leads to microvascular (retinopathy, nephropathy, neuropathy) and macrovascular (cardiovascular disease) complications

KEY CONCEPTS: DYSREGULATION OF GLUCOSE HOMEOSTASIS IN DIABETES MELLITUS

Causes, Clinical Signs and Symptoms of Applied Aspects of Diabetes Mellitus (Table 2). [1 - 3]

Table 2. Key aspects of diabetes.

Aspect	Type 1 Diabetes Mellitus (T1DM)	Type 2 Diabetes Mellitus (T2DM)
Pathogenesis	- Autoimmune destruction of pancreatic β-cells	- Insulin resistance in peripheral tissues (muscle, fat, liver)
	- Absolute insulin deficiency	- Progressive β-cell dysfunction
Insulin Levels	- Severely reduced or absent	- Initially normal or elevated; decreases over time
Glucagon Levels	- Elevated due to lack of insulin inhibition	- Often elevated or inadequately suppressed
Blood Glucose Regulation	- Hyperglycemia due to lack of insulin-mediated glucose uptake	- Hyperglycemia due to insulin resistance and impaired insulin secretion
Glycogen Metabolism	- Decreased glycogen synthesis in the liver and muscle	- Impaired glycogen synthesis and increased glycogen breakdown
Gluconeogenesis	- Increased hepatic gluconeogenesis due to unopposed glucagon action	- Increased hepatic gluconeogenesis despite hyperinsulinemia
Lipid Metabolism	- Increased lipolysis leading to elevated free fatty acids	- Increased lipolysis and decreased lipid storage due to insulin resistance
	- Ketogenesis leading to ketoacidosis	- Mild ketogenesis, but usually not severe due to residual insulin activity
Protein Metabolism	- Increased proteolysis due to lack of insulin	- Increased proteolysis and reduced protein synthesis due to insulin resistance

(Table 2) cont.....

Aspect	Type 1 Diabetes Mellitus (T1DM)	Type 2 Diabetes Mellitus (T2DM)
Hormonal Dysregulation	- Lack of insulin leads to unregulated glucagon, cortisol, and growth hormone action	- Insulin resistance leads to compensatory hyperinsulinemia, dysregulated glucagon
Clinical Symptoms	- Polyuria, polydipsia, polyphagia, weight loss, fatigue, and blurred vision	- Polyuria, polydipsia, polyphagia, fatigue, and frequent infections
Acute Complications	- Diabetic Ketoacidosis (DKA) due to severe insulin deficiency	- Hyperosmolar Hyperglycemic State (HHS) due to severe hyperglycemia and dehydration
Chronic Complications	- Microvascular: Retinopathy, nephropathy, neuropathy	- Microvascular: Retinopathy, nephropathy, neuropathy
	- Macrovascular: Cardiovascular disease, cerebrovascular disease, peripheral artery disease	- Macrovascular: Cardiovascular disease, cerebrovascular disease, peripheral artery disease
Diagnostic Markers	- Elevated fasting blood glucose, HbA1c, and autoantibodies (*e.g.*, GAD, IA-2, ZnT8)	- Elevated fasting blood glucose, HbA1c, and markers of insulin resistance (*e.g.*, HOMA-IR)
Management	- Insulin therapy, blood glucose monitoring, dietary management	- Lifestyle modification, oral hypoglycemic agents, insulin therapy if needed

KEY CONCEPTS: ETIOLOGY OF TYPE I, TYPE 2, AND GESTATIONAL DIABETES (TABLE 3)

Table 3. Etiology of diabetes [1 - 5].

Type	Etiology	Genetic Factors	Environmental/Lifestyle Factors	Characteristics
Type 1 Diabetes	Autoimmune destruction of β-cells in the pancreas	HLA-DR3, HLA-DR4, other genetic loci	Viral infections (*e.g.*, enteroviruses), early exposure to cow's milk, dietary components	Typically abrupt onset, most common in childhood/adolescence, absolute insulin deficiency
Type 2 Diabetes	Insulin resistance and progressive β-cell dysfunction	Numerous genetic loci linked to increased risk	Obesity, physical inactivity, poor diet, central obesity, aging, certain medications	Gradual onset, often asymptomatic for years, strongly linked to lifestyle factors, relative insulin deficiency

(Table 3) cont.....

Type	Etiology	Genetic Factors	Environmental/Lifestyle Factors	Characteristics
Gestational Diabetes	Insulin resistance during pregnancy surpasses β-cell capacity to produce insulin	Family history of diabetes, certain ethnic backgrounds	Obesity, advanced maternal age, history of GDM, or high birth weight baby	Develops during pregnancy, usually resolves after childbirth, increases the risk of T2DM later in life for the mother/child

KEY CONCEPTS: DIABETES, COMPLICATIONS AND CLINICAL MANAGEMENT (TABLE 4) [1, 5,6]

Table 4. Diabetes, complications, and applied aspects.

Topic	Key Concepts	Applied Aspects
Complications of Diabetes (Microvascular)	- Microvascular complications include retinopathy, nephropathy, and neuropathy. - Caused by chronic hyperglycemia leading to damage of small blood vessels.	- Regular screening for diabetic retinopathy(fundoscopy), urine albumin for nephropathy, and neurological exams.
Complications of Diabetes (Macrovascular)	- Macrovascular complications include cardiovascular disease, stroke, and peripheral artery disease. - Accelerated atherosclerosis due to chronic hyperglycemia and dyslipidemia.	- Monitor for cardiovascular risk factors: lipid profile, blood pressure, ECG. - Use of statins, ACE inhibitors, and antiplatelet agents.
Diabetic Ketoacidosis (DKA)	- A life-threatening condition primarily seen in Type 1 diabetes. - Characterized by hyperglycemia, ketonemia, and metabolic acidosis. - Triggers: infection, insulin omission, and stress.	- Lab tests: serum glucose, ketones, Arterial Blood Gases (ABG). - Management: IV fluids, insulin, electrolyte correction (especially K^+).
HbA1c Standardization	- Glycated Hemoglobin (HbA1c) reflects average blood glucose over 2-3 months. - Standardized as a primary diagnostic and monitoring tool for diabetes.	- HbA1c ≥ 6.5% is diagnostic for diabetes. - Target: < 7% for good control in most patients. - Adjust for conditions like anemia, hemoglobinopathies.

(Table 4) cont.....

Topic	Key Concepts	Applied Aspects
Guidelines for Screening, Diagnosis, and Monitoring of Diabetes	- Screening recommended for high-risk populations (obesity, family history, hypertension). - Diagnosis *via* fasting plasma glucose (≥ 126 mg/dL), HbA1c (≥ 6.5%), or OGTT.	- Annual screening for complications: retinopathy, nephropathy (urine microalbumin), neuropathy. - Regular HbA1c testing every 3-6 months.
Non-Ketotic Hyperosmolar Coma (HHS)	- Primarily affects Type 2 diabetes. - Severe hyperglycemia, hyperosmolality, without significant ketosis. - Often precipitated by infections or other illnesses.	- Lab tests: serum glucose, osmolality, electrolytes, renal function. - Management: IV fluids, electrolyte correction, and insulin therapy.
Self-Monitoring of Blood Glucose (SMBG)	- Key for day-to-day management, especially in patients on insulin. - Helps detect hypoglycemia and adjust therapy.	- Frequency depends on diabetes type and treatment (*e.g.*, multiple times daily for insulin users). - Use of glucometers and Continuous Glucose Monitors (CGM).

BIOCHEMICAL BASIS, RELEVANT LAB TESTS, INTERPRETATION OF RESULTS, AND APPLIED ASPECTS

Laboratory Investigations that are Important in the Detection, Diagnosis, and Management of Complications (Table 5). [2, 6]

Table 5. Lab investigations in diabetes.

Condition	Biochemical Basis	Relevant Lab Tests	Interpretation of Results	Applied Aspects
Diabetic Ketoacidosis (DKA)	- Insulin deficiency leading to lipolysis and ketogenesis. - Results in hyperglycemia, ketonemia, and metabolic acidosis.	- Serum glucose, ketones (β-hydroxybutyrate), Arterial Blood Gases (ABG), electrolytes, and anion gap.	- High glucose (>250 mg/dL), low bicarbonate, high anion gap, ketones positive, acidic pH. - Decreased K^+ despite normal serum K^+.	- Management involves IV fluids, insulin infusion, and electrolyte replacement (K^+ correction is critical).

(Table 5) cont.....

Condition	Biochemical Basis	Relevant Lab Tests	Interpretation of Results	Applied Aspects
Gestational Diabetes	- Temporary insulin resistance during pregnancy due to placental hormones like human placental lactogen.	- Oral Glucose Tolerance Test (OGTT), fasting plasma glucose, HbA1c, postprandial glucose.	- OGTT 1-hour post-load glucose >180 mg/dL. - Fasting glucose >92 mg/dL. HbA1c ≥6.5% indicates diabetes.	- Monitoring glucose during pregnancy to avoid fetal complications. - Lifestyle management or insulin therapy.
Hypoglycemia	- Excessive insulin or lack of glucose availability leading to low blood sugar. - Common in insulin-treated diabetics.	- Serum glucose, C-peptide, insulin, and glucagon stimulation test.	- Glucose <70 mg/dL. - High insulin with low C-peptide suggests exogenous insulin use.	- Immediate treatment: oral glucose or IV dextrose. - Adjust insulin or medication dose to prevent future episodes.
Metabolic Syndrome	- Cluster of conditions: insulin resistance, hyperglycemia, hypertension, dyslipidemia, and central obesity.	- Fasting glucose, HbA1c, lipid profile (triglycerides, HDL), blood pressure, and waist circumference.	- Fasting glucose >100 mg/dL. - High triglycerides (>150 mg/dL), low HDL (<40 mg/dL for men, <50 mg/dL for women).	- Management focuses on lifestyle changes: diet, exercise, and pharmacotherapy to treat dyslipidemia and hypertension.
Non-Ketotic Hyperosmolar Coma (HHS)	- Severe hyperglycemia without significant ketosis, seen in Type 2 diabetes. - Results in hyperosmolality and dehydration.	- Serum glucose, osmolality, electrolytes, renal function (creatinine), and ABG.	- Serum glucose >600 mg/dL. - Elevated osmolality (>320 mOsm/kg). - No significant ketosis or acidosis.	- Treatment involves aggressive fluid replacement, insulin, and electrolyte management. - Prevention focuses on better glucose control.
Pre-diabetes/Impaired Glucose Tolerance (IGT)	- Early stage of insulin resistance or beta-cell dysfunction.	- Fasting glucose, HbA1c, OGTT.	- Fasting glucose 100-125 mg/dL. - HbA1c 5.7%-6.4%. - OGTT 2-hour plasma glucose 140-199 mg/dL.	- Lifestyle modification to delay or prevent progression to Type 2 diabetes. - Dietary changes and physical activity are recommended.

(Table 5) cont.....

Condition	Biochemical Basis	Relevant Lab Tests	Interpretation of Results	Applied Aspects
Type 1 Diabetes	- Autoimmune destruction of pancreatic beta cells, leading to insulin deficiency.	- Fasting glucose, HbA1c, C-peptide, autoantibodies (*e.g.*, GAD, IA-2, ZnT8).	- Fasting glucose ≥126 mg/dL. - HbA1c ≥6.5%. - Low C-peptide, positive autoantibodies (*e.g.*, GAD).	- Lifelong insulin therapy. - Regular monitoring of glucose and HbA1c to prevent complications (*e.g.*, DKA).
Type 2 Diabetes	- Insulin resistance with a relative insulin deficiency. - Associated with obesity, inactivity, and genetics.	- Fasting glucose, HbA1c, OGTT, and C-peptide.	- Fasting glucose ≥126 mg/dL. - HbA1c ≥6.5%. - Normal or high C-peptide levels, absence of autoantibodies.	- Initial management with diet and exercise. - Pharmacotherapy includes metformin, SGLT2 inhibitors, and insulin in advanced cases.

RELEVANT LABORATORY TESTS, ANALYTICS METHODS AVAILABLE AND APPLIED ASPECTS (TABLE 5, FIG. (1-3): [2]

Table 6. Analytical methods for diabetes investigations.

Test	Purpose/Detection	Analytics Methods	Applied Aspects
Anti-Glutamic Acid Decarboxylase (GAD) antibodies	Detects autoantibodies in Type 1 diabetes.	ELISA, RIA (radioimmunoassay)	Positive results indicate autoimmune destruction of beta cells; used in the diagnosis of Type 1 diabetes or latent autoimmune diabetes.
Anti-insulin antibodies	Identifies antibodies against exogenous insulin.	ELISA, RIA	Used to identify immune response to exogenous insulin in patients receiving insulin; helps adjust insulin therapy.
Blood gas and hydrogen ion measurements	Measures acid-base balance and pH in the blood (*e.g.*, DKA, HHS).	Arterial Blood Gas (ABG), ion-selective electrodes	Useful in monitoring diabetic ketoacidosis and other metabolic disorders; detects metabolic acidosis/alkalosis.
C-peptide	Assesses endogenous insulin production.	Immunoassay, RIA	Differentiates between Type 1 and Type 2 diabetes; low C-peptide in Type 1, normal/high in Type 2; monitors insulinomas.

(Table 6) cont.....

Test	Purpose/Detection	Analytics Methods	Applied Aspects
Fructosamine and glycated proteins	Measures short-term (2-3 weeks) average blood glucose.	Colorimetric assay, HPLC	Used for glycemic control in situations where HbA1c is not reliable (*e.g.*, hemoglobinopathies, recent transfusion).
Glucagon	Measures pancreatic function and glucose regulation.	RIA, ELISA	Increased in glucagonoma, reduced in hypoglycemia; monitors insulin and glucagon interplay.
Glucose	Measures blood glucose levels for diagnosis and monitoring of diabetes.	Glucose oxidase method, hexokinase method	Key diagnostic test for diabetes; used in routine monitoring, diagnosis (*e.g.*, fasting glucose, random glucose levels). (Fig. 1)
Glucose Tolerance Test (OGTT)	Assesses the body's response to an oral glucose load for diagnosing diabetes.	Blood glucose measurements before/after glucose intake	Diagnostic for gestational diabetes and Type 2 diabetes (2-hour plasma glucose ≥ 200 mg/dL indicates diabetes).
HbA1c	Reflects average blood glucose over 2-3 months.	HPLC, immunoassay, capillary electrophoresis	Key test for long-term glucose control in diabetic patients; HbA1c ≥ 6.5% diagnostic for diabetes. (Fig. 2)
Insulin	Measures circulating insulin levels.	Immunoassay, RIA	Helps diagnose insulinomas, differentiate Type 1 vs. Type 2 diabetes; high levels in insulin resistance. (Fig. 3)
Insulinoma antigen 2 (IA-2)	Autoantibodies linked to Type 1 diabetes (autoimmune destruction of beta cells).	ELISA, RIA	IA-2 antibodies are used to support the diagnosis of Type 1 diabetes; often assessed alongside GAD antibodies.
Ketones (β-hydroxy butyrate)	Measures ketone bodies, particularly in Diabetic Ketoacidosis (DKA).	Enzymatic assay, blood/urine dipstick	Elevated in DKA and severe insulin deficiency; used for the diagnosis and monitoring of DKA.
Microalbumin	Detects early-stage diabetic nephropathy.	Immunoassay, turbidimetry, HPLC	Early marker of kidney damage in diabetic patients; recommended annual test for diabetes management.
Zinc Transporter 8 (ZnT8)	Autoantibodies associated with Type 1 diabetes.	ELISA, RIA	Helps confirm the diagnosis of Type 1 diabetes; present in newly diagnosed patients and used in predictive testing.

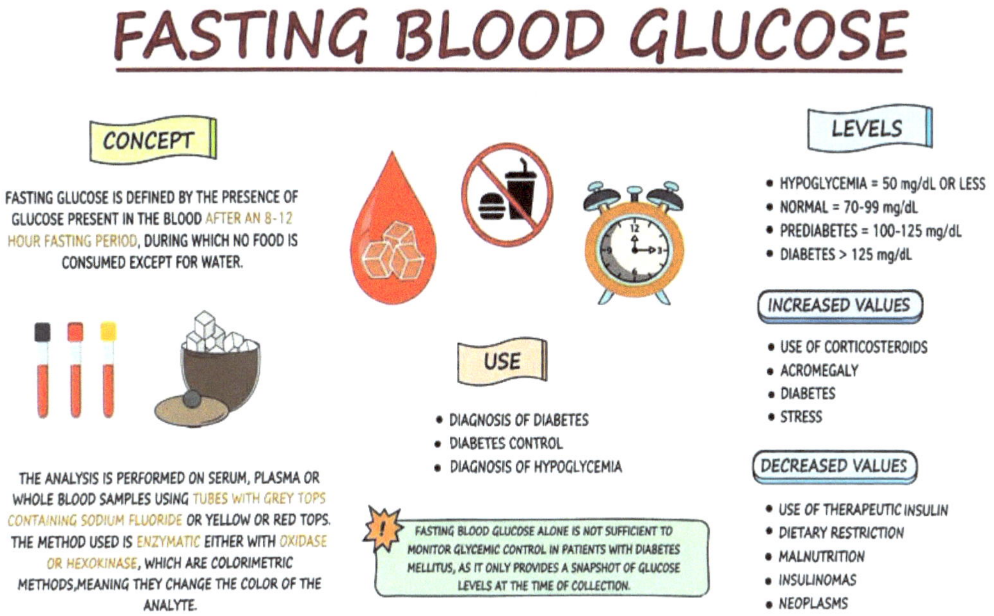

Fig. (1). Fasting blood glucose.

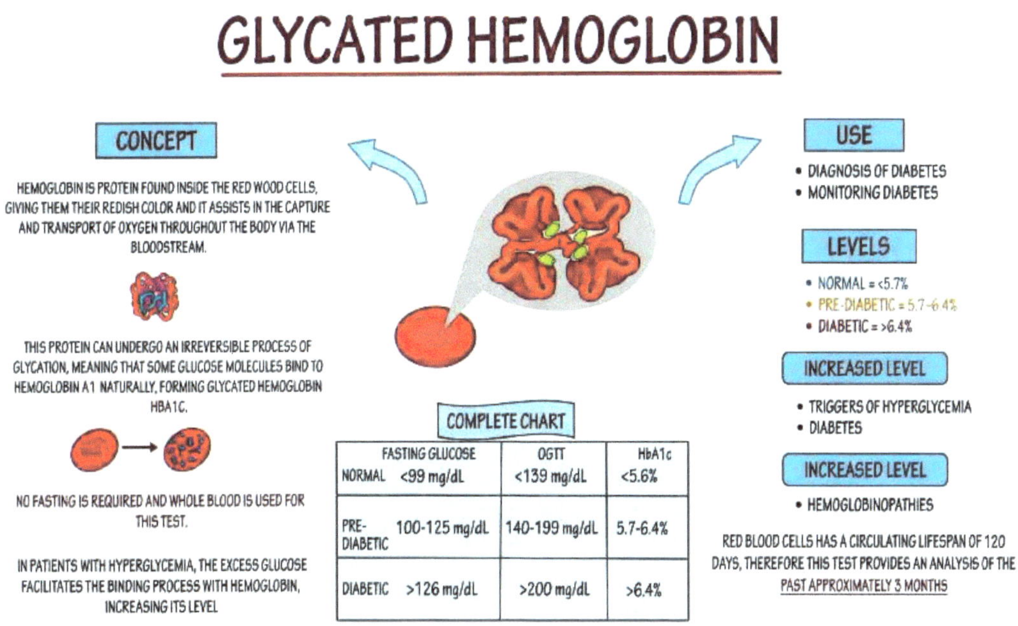

Fig. (2). Other types of tests.

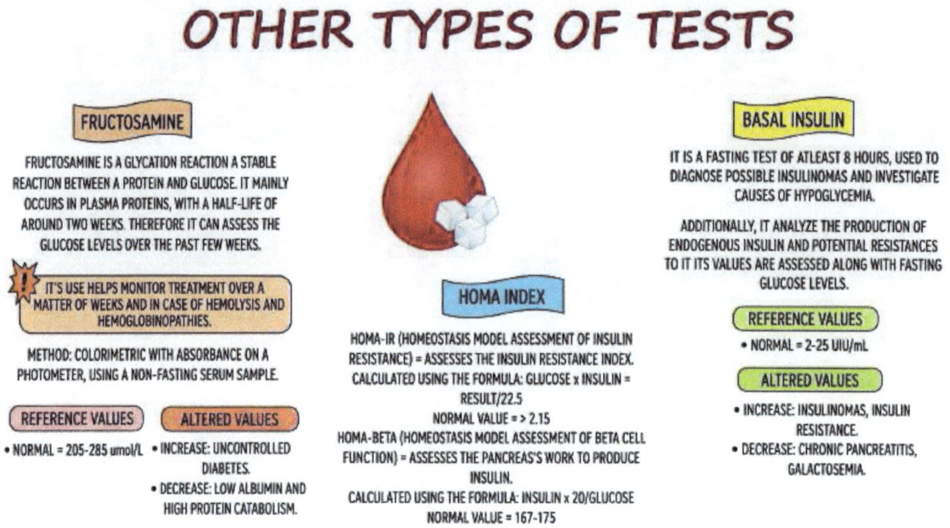

Fig. (3). Glycated haemoglobin.

Summary

Diabetes mellitus is primarily classified as Type 1 (T1DM), Type 2 (T2DM), and Gestational Diabetes (GDM). Early diagnosis and aggressive management of hyperglycemia are essential to reduce complications like Diabetic Ketoacidosis (DKA), hypoglycemia, and both microvascular and macrovascular complications.

Exercises

- What are the primary biochemical markers used in the diagnosis of Type 1 and Type 2 diabetes?
- Explain the roles of fasting glucose, HbA1c, and OGTT in diagnosing diabetes.
- Differentiate between Diabetic Ketoacidosis (DKA) and Hyperosmolar Hyperglycemic State (HHS) in terms of biochemical markers and clinical presentation.
- Discuss the significance of HbA1c in long-term glycemic control.
- Describe the laboratory tests used to differentiate between Type 1 and Type 2 diabetes.
- Explain the guidelines for screening gestational diabetes mellitus.
- Discuss the role and limitations of Continuous Glucose Monitoring (CGM) in managing diabetes.
- What are the biochemical indicators of diabetic nephropathy?

CONCLUSION

Effective management of diabetes mellitus relies heavily on accurate and timely biochemical diagnosis and monitoring. Education on self-monitoring and regular screening for complications is also a key aspect in the management of diabetic patients. Continued research into new biomarkers and personalized treatment approaches will further improve patient outcomes.

REFERENCES

[1] Classification and Diagnosis of Diabetes: *Standards of Medical Care in Diabetes—2021. Diabetes Care,* **2021**, *44* Suppl. 1, S15-S33. [http://dx.doi.org/10.2337/dc21-S002] [PMID: 33298413]

[2] Zhou, B.; Sheffer, K.E.; Bennett, J.E.; Gregg, E.W.; Danaei, G.; Singleton, R.K.; Shaw, J.E.; Mishra, A.; Lhoste, V.P.F.; Carrillo-Larco, R.M.; Kengne, A.P.; Phelps, N.H.; Heap, R.A.; Rayner, A.W.; Stevens, G.A.; Paciorek, C.J.; Riley, L.M.; Cowan, M.J.; Savin, S.; Vander Hoorn, S.; Lu, Y.; Pavkov, M.E.; Imperatore, G.; Aguilar-Salinas, C.A.; Ahmad, N.A.; Anjana, R.M.; Davletov, K.; Farzadfar, F.; González-Villalpando, C.; Khang, Y-H.; Kim, H.C.; Laatikainen, T.; Laxmaiah, A.; Mbanya, J.C.N.; Narayan, K.M.V.; Ramachandran, A.; Wade, A.N.; Zdrojewski, T.; Abbasi-Kangevari, M.; Rahim, H.F.A.; Abu-Rmeileh, N.M.; Adambekov, S.; Adams, R.J.; Aekplakorn, W.; Agdeppa, I.A.; Aghazadeh-Attari, J.; Agyemang, C.; Ahmadi, A.; Ahmadi, N.; Ahmadi, N.; Ahmed, S.H.; Ajlouni, K.; Al-Hinai, H.; Al-Lahou, B.; Al-Lawati, J.A.; Asfoor, D.A.; Al Qaoud, N.M.; Alarouj, M.; AlBuhairan, F.; AlDhukair, S.; Aldwairji, M.A.; Ali, M.M.; Alinezhad, F.; Alkandari, A.; Alomirah, H.F.; Aly, E.; Amarapurkar, D.N.; Andersen, L.B.; Anderssen, S.A.; Andrade, D.S.; Ansari-Moghaddam, A.; Aounallah-Skhiri, H.; Aris, T.; Arlappa, N.; Aryal, K.K.; Assah, F.K.; Assembekov, B.; Auvinen, J.; Avdičová, M.; Azad, K.; Azimi-Nezhad, M.; Azizi, F.; Bacopoulou, F.; Balakrishna, N.; Bamoshmoosh, M.; Banach, M.; Bandosz, P.; Banegas, J.R.; Barbagallo, C.M.; Barceló, A.; Baretić, M.; Barrera, L.; Basit, A.; Batieha, A.M.; Batista, A.P.; Baur, L.A.; Belavendra, A.; Ben Romdhane, H.; Benet, M.; Berkinbayev, S.; Bernabe-Ortiz, A.; Berrios Carrasola, X.; Bettiol, H.; Beybey, A.F.; Bhargava, S.K.; Bika Lele, E.C.; Bikbov, M.M.; Bista, B.; Bjerregaard, P.; Bjertness, E.; Bjertness, M.B.; Björkelund, C.; Bloch, K.V.; Blokstra, A.; Bo, S.; Bobak, M.; Boggia, J.G.; Bonaccio, M.; Bonilla-Vargas, A.; Borghs, H.; Bovet, P.; Brajkovich, I.; Brenner, H.; Brewster, L.M.; Brian, G.R.; Briceño, Y.; Brito, M.; Bugge, A.; Buntinx, F.; Cabrera de León, A.; Caixeta, R.B.; Can, G.; Cândido, A.P.C.; Capanzana, M.V.; Čapková, N.; Capuano, E.; Capuano, R.; Capuano, V.; Cardoso, V.C.; Carlsson, A.C.; Casanueva, F.F.; Censi, L.; Cervantes-Loaiza, M.; Chamnan, P.; Chamukuttan, S.; Chan, Q.; Charchar, F.J.; Chaturvedi, N.; Chen, H.; Cheraghian, B.; Chirlaque, M-D.; Chudek, J.; Cifkova, R.; Cirillo, M.; Claessens, F.; Cohen, E.; Concin, H.; Cooper, C.; Costanzo, S.; Cowell, C.; Crujeiras, A.B.; Cruz, J.J.; Cureau, F.V.; Cuschieri, S.; D'Arrigo, G.; d'Orsi, E.; Dallongeville, J.; Damasceno, A.; Dastgiri, S.; De Curtis, A.; de Gaetano, G.; De Henauw, S.; Deepa, M.; DeGennaro, V., Jr; Demarest, S.; Dennison, E.; Deschamps, V.; Dhimal, M.; Dika, Z.; Djalalinia, S.; Donfrancesco, C.; Dong, G.; Dorobantu, M.; Dörr, M.; Dragano, N.; Drygas, W.; Du, Y.; Duante, C.A.; Duboz, P.; Dushpanova, A.; Dziankowska-Zaborszczyk, E.; Ebrahimi, N.; Eddie, R.; Eftekhar, E.; Efthymiou, V.; Egbagbe, E.E.; Eghtesad, S.; El-Khateeb, M.; El Ati, J.; Eldemire-Shearer, D.; Elosua, R.; Enang, O.; Erasmus, R.T.; Erbel, R.; Erem, C.; Ergor, G.; Eriksen, L.; Eriksson, J.G.; Esmaeili, A.; Evans, R.G.; Fakhradiyev, I.; Fall, C.H.; Faramarzi, E.; Farjam, M.; Farzi, Y.; Fattahi, M.R.; Fawwad, A.; Felix-Redondo, F.J.; Ferguson, T.S.; Fernández-Bergés, D.; Ferrari, M.; Ferreccio, C.; Ferreira, H.S.; Ferrer, E.; Feskens, E.J.M.; Flood, D.; Forsner, M.; Fosse, S.; Fottrell, E.F.; Fouad, H.M.; Francis, D.K.; Frontera, G.; Furusawa, T.; Gaciong, Z.; Garnett, S.P.; Gasull, M.; Gazzinelli, A.; Gehring, U.; Ghaderi, E.; Ghamari, S-H.; Ghanbari, A.; Ghasemi, E.; Gheorghe-Fronea, O-F.; Ghimire, A.; Gialluisi, A.; Giampaoli, S.; Gianfagna, F.; Gill, T.K.; Gironella, G.; Giwercman, A.; Goltzman, D.; Gomula, A.; Gonçalves, H.; Gonçalves, M.; Gonzalez-Chica, D.A.; Gonzalez-Gross, M.; González-Rivas, J.P.; González-Villalpando, M-E.; Gonzalez, A.R.; Gottrand, F.; Grafnetter, D.; Grodzicki, T.; Grøntved, A.; Guerrero, R.; Gujral, U.P.; Gupta, R.; Gutierrez, L.; Gwee, X.;

Haghshenas, R.; Hakimi, H.; Hambleton, I.R.; Hamzeh, B.; Hanekom, W.A.; Hange, D.; Hantunen, S.; Hao, J.; Hari Kumar, R.; Harooni, J.; Hashemi-Shahri, S.M.; Hata, J.; Heidemann, C.; Henrique, R.S.; Herrala, S.; Herzig, K-H.; Heshmat, R.; Ho, S.Y.; Holdsworth, M.; Homayounfar, R.; Hopman, W.M.; Horimoto, A.R.V.R.; Hormiga, C.; Horta, B.L.; Houti, L.; Howitt, C.; Htay, T.T.; Htet, A.S.; Htike, M.M.T.; Huerta, J.M.; Huhtaniemi, I.T.; Huisman, M.; Husseini, A.; Huybrechts, I.; Iacoviello, L.; Iakupova, E.M.; Iannone, A.G.; Ibrahim Wong, N.; Ijoma, C.; Irazola, V.E.; Ishida, T.; Isiguzo, G.C.; Islam, S.M.S.; Islek, D.; Ittermann, T.; Iwasaki, M.; Jääskeläinen, T.; Jacobs, J.M.; Jaddou, H.Y.; Jadoul, M.; Jallow, B.; James, K.; Jamil, K.M.; Janus, E.; Jarvelin, M-R.; Jasienska, G.; Jelaković, A.; Jelaković, B.; Jennings, G.; Jha, A.K.; Jimenez, R.O.; Jöckel, K-H.; Jokelainen, J.J.; Jonas, J.B.; Joshi, P.; Josipović, J.; Joukar, F.; Jóźwiak, J.; Kafatos, A.; Kajantie, E.O.; Kalmatayeva, Z.; Karki, K.B.; Katibeh, M.; Kauhanen, J.; Kazakbaeva, G.M.; Kaze, F.F.; Ke, C.; Keinänen-Kiukaanniemi, S.; Kelishadi, R.; Keramati, M.; Kersting, M.; Khader, Y.S.; Khaledifar, A.; Khalili, D.; Kheiri, B.; Kheradmand, M.; Khosravi, A.; Kiechl-Kohlendorfer, U.; Kiechl, S.J.; Kiechl, S.; Kingston, A.; Klakk, H.; Klanova, J.; Knoflach, M.; Kolsteren, P.; König, J.; Korpelainen, R.; Korrovits, P.; Kos, J.; Koskinen, S.; Kowlessur, S.; Koziel, S.; Kriemler, S.; Kristensen, P.L.; Kromhout, D.; Kubinova, R.; Kujala, U.M.; Kulimbet, M.; Kurjata, P.; Kyobutungi, C.; La, Q.N.; Labadarios, D.; Lachat, C.; Laid, Y.; Lall, L.; Lankila, T.; Lanska, V.; Lappas, G.; Larijani, B.; Latt, T.S.; Laurenzi, M.; Lehmann, N.; Lehtimäki, T.; Lemogoum, D.; Leung, G.M.; Li, Y.; Lima-Costa, M.F.; Lin, H-H.; Lind, L.; Lissner, L.; Liu, X.; Lopez-Garcia, E.; Lopez, T.; Lozano, J.E.; Luksiene, D.; Lundqvist, A.; Lunet, N.; Lustigová, M.; Machado-Coelho, G.L.L.; Machado-Rodrigues, A.M.; Macia, E.; Macieira, L.M.; Madar, A.A.; Maestre, G.E.; Maggi, S.; Magliano, D.J.; Magriplis, E.; Mahasampath, G.; Maire, B.; Makdisse, M.; Malekpour, M-R.; Malekzadeh, F.; Malekzadeh, R.; Mallikharjuna Rao, K.; Malyutina, S.; Maniego, L.V.; Manios, Y.; Mannix, M.I.; Mansour-Ghanaei, F.; Manzato, E.; Margozzini, P.; Mariño, J.; Marques, L.P.; Martorell, R.; Mascarenhas, L.P.; Masinaei, M.; Mathiesen, E.B.; Matsha, T.E.; Mc Donald Posso, A.J.; McFarlane, S.R.; McGarvey, S.T.; Mediene Benchekor, S.; Mehlig, K.; Mehrparvar, A.H.; Melgarejo, J.D.; Méndez, F.; Menezes, A.M.B.; Mereke, A.; Meshram, I.I.; Meto, D.T.; Minderico, C.S.; Mini, G.K.; Miquel, J.F.; Miranda, J.J.; Mirjalili, M.R.; Modesti, P.A.; Moghaddam, S.S.; Mohamed, M.K.; Mohammad, K.; Mohammadi, M.R.; Mohammadi, Z.; Mohammadifard, N.; Mohammadpourhodki, R.; Mohan, V.; Mohd Yusoff, M.F.; Mohebbi, I.; Møller, N.C.; Molnár, D.; Momenan, A.; Mondo, C.K.; Montenegro Mendoza, R.A.; Monterrubio-Flores, E.; Moosazadeh, M.; Moradpour, F.; Morejon, A.; Moreno, L.A.; Morgan, K.; Morin, S.N.; Moslem, A.; Mosquera, M.; Mossakowska, M.; Mostafa, A.; Mostafavi, S-A.; Motlagh, M.E.; Motta, J.; Msyamboza, K.P.; Mu, T.T.; Muiesan, M.L.; Mursu, J.; Musa, K.I.; Mustafa, N.; Muyer, M.T.M.C.; Nabipour, I.; Nagel, G.; Naidu, B.M.; Najafi, F.; Námešná, J.; Nangia, V.B.; Naseri, T.; Neelapaichit, N.; Nejatizadeh, A.; Nenko, I.; Nervi, F.; Ng, T.P.; Nguyen, C.T.; Nguyen, Q.N.; Ni, M.Y.; Nie, P.; Nieto-Martínez, R.E.; Ninomiya, T.; Noale, M.; Noboa, O.A.; Noto, D.; Nsour, M.A.; Nuhoğlu, I.; O'Neill, T.W.; Odili, A.N.; Oh, K.; Ohtsuka, R.; Omar, M.A.; Onat, A.; Ong, S.K.; Onodugo, O.; Ordunez, P.; Ornelas, R.; Ortiz, P.J.; Osmond, C.; Ostovar, A.; Otero, J.A.; Ottendahl, C.B.; Otu, A.; Owusu-Dabo, E.; Palmieri, L.; Pan, W-H.; Panda-Jonas, S.; Panza, F.; Paoli, M.; Park, S.; Parsaeian, M.; Patel, N.D.; Pechlaner, R.; Pećin, I.; Pedro, J.M.; Peixoto, S.V.; Peltonen, M.; Pereira, A.C.; Pessôa dos Prazeres, T.M.; Peykari, N.; Phall, M.C.; Pham, S.T.; Phan, H.H.; Pichardo, R.N.; Pikhart, H.; Pilav, A.; Piler, P.; Pitakaka, F.; Piwonska, A.; Pizarro, A.N.; Plans-Rubió, P.; Plata, S.; Porta, M.; Poudyal, A.; Pourfarzi, F.; Pourshams, A.; Poustchi, H.; Pradeepa, R.; Providencia, R.; Puder, J.J.; Puhakka, S.; Punab, M.; Qorbani, M.; Quintana, H.K.; Quoc Bao, T.; Rahimikazerooni, S.; Raitakari, O.; Ramirez-Zea, M.; Ramke, J.; Ramos, R.; Rampal, L.; Rampal, S.; Rangel Reina, D.A.; Rashidi, M-M.; Redon, J.; Renner, J.D.P.; Reuter, C.P.; Revilla, L.; Rezaei, N.; Rezaianzadeh, A.; Rigo, F.; Roa, R.G.; Robinson, L.; Rodríguez-Artalejo, F.; Rodriguez-Perez, M.C.; Rodríguez-Villamizar, L.A.; Rodríguez, A.Y.; Roggenbuck, U.; Rohloff, P.; Romeo, E.L.; Rosengren, A.; Rubinstein, A.; Rust, P.; Rutkowski, M.; Sabbaghi, H.; Sachdev, H.S.; Sadjadi, A.; Safarpour, A.R.; Safi, S.; Safiri, S.; Saghi, M.H.; Saidi, O.; Saki, N.; Šalaj, S.; Salanave, B.; Salonen, J.T.; Salvetti, M.; Sánchez-Abanto, J.; Santos, D.A.; Santos, L.C.; Santos, M.P.; Santos, T.R.; Saramies, J.L.; Sardinha, L.B.; Sarrafzadegan, N.; Saum, K-U.; Sbaraini, M.; Scazufca, M.; Schaan, B.D.; Scheidt-Nave, C.; Schipf, S.; Schmidt, C.O.; Schöttker, B.; Schramm, S.; Sebert, S.; Sedaghattalab, M.; Sein, A.A.; Sepanlou, S.G.; Sewpaul, R.; Shamah-Levy, T.; Shamshirgaran, S.M.; Sharafkhah, M.; Sharma, S.K.; Sharman, A.; Shayanrad, A.; Shayesteh, A.A.; Shimizu-Furusawa, H.; Shiri, R.; Shrestha, N.; Si-Ramlee, K.; Silva, D.A.S.;

Simon, M.; Simons, J.; Simons, L.A.; Sjöström, M.; Slowikowska-Hilczer, J.; Slusarczyk, P.; Smeeth, L.; Sobngwi, E.; Söderberg, S.; Soemantri, A.; Sofat, R.; Solfrizzi, V.; Somi, M.H.; Soumaré, A.; Sousa-Poza, A.; Sparrenberger, K.; Staessen, J.A.; Stavreski, B.; Steene-Johannessen, J.; Stehle, P.; Stein, A.D.; Stessman, J.; Stokwiszewski, J.; Stronks, K.; Suarez-Ortegón, M.F.; Suebsamran, P.; Sundström, J.; Suriyawongpaisal, P.; Sylva, R.C.; Szklo, M.; Tamosiunas, A.; Tarawneh, M.R.; Tarqui-Mamani, C.B.; Taylor, A.; Taylor, J.; Tello, T.; Thankappan, K.R.; Theobald, H.; Theodoridis, X.; Thomas, N.; Thrift, A.G.; Timmermans, E.J.; Tjandrarini, D.H.; Tolonen, H.K.; Tolstrup, J.S.; Tomaszewski, M.; Topbas, M.; Torres-Collado, L.; Traissac, P.; Triantafyllou, A.; Tuitele, J.; Tuliakova, A.M.; Tulloch-Reid, M.K.; Tuomainen, T-P.; Tzala, E.; Tzourio, C.; Ueda, P.; Ugel, E.; Ukoli, F.A.M.; Ulmer, H.; Uusitalo, H.M.T.; Valdivia, G.; van den Born, B-J.; Van der Heyden, J.; Van Minh, H.; van Rossem, L.; Van Schoor, N.M.; van Valkengoed, I.G.M.; van Zutphen, E.M.; Vanderschueren, D.; Vanuzzo, D.; Vasan, S.K.; Vega, T.; Velasquez-Melendez, G.; Verstraeten, R.; Viet, L.; Villalpando, S.; Vioque, J.; Virtanen, J.K.; Viswanathan, B.; Voutilainen, A.; Wan Bebakar, W.M.; Wan Mohamud, W.N.; Wang, C.; Wang, N.; Wang, Q.; Wang, Y.X.; Wang, Y-W.; Wannamethee, S.G.; Webster-Kerr, K.; Wedderkopp, N.; Wei, W.; Westbury, L.D.; Whincup, P.H.; Widhalm, K.; Widyahening, I.S.; Więcek, A.; Wilks, R.J.; Willeit, J.; Willeit, P.; Wilsgaard, T.; Wojtyniak, B.; Wong, A.; Wong, E.B.; Woodward, M.; Wu, F.C.; Xu, H.; Xu, L.; Yaacob, N.A.; Yan, L.; Yan, W.; Yoosefi, M.; Yoshihara, A.; Younger-Coleman, N.O.; Yu, Y-L.; Yu, Y.; Yusoff, A.F.; Zainuddin, A.A.; Zamani, F.; Zambon, S.; Zampelas, A.; Zaw, K.K.; Zeljkovic Vrkic, T.; Zeng, Y.; Zhang, Z-Y.; Zholdin, B.; Zimmet, P.; Zitt, E.; Zoghlami, N.; Zuñiga Cisneros, J.; Ezzati, M. Global variation in diabetes diagnosis and prevalence based on fasting glucose and hemoglobin A1c. *Nat. Med.,* **2023**, *29*(11), 2885-2901.
[http://dx.doi.org/10.1038/s41591-023-02610-2] [PMID: 37946056]

[3] ElSayed, N.A.; Aleppo, G.; Aroda, V.R.; Bannuru, R.R.; Brown, F.M.; Bruemmer, D.; Collins, B.S.; Gaglia, J.L.; Hilliard, M.E.; Isaacs, D.; Johnson, E.L.; Kahan, S.; Khunti, K.; Leon, J.; Lyons, S.K.; Perry, M.L.; Prahalad, P.; Pratley, R.E.; Seley, J.J.; Stanton, R.C.; Gabbay, R.A. 2. Classification and Diagnosis of Diabetes: *Standards of Care in Diabetes—2023. Diabetes Care,* **2023**, *46* Suppl. 1, S19-S40.
[http://dx.doi.org/10.2337/dc23-S002] [PMID: 36507649]

[4] Park, Y.; Ko, K.S.; Rhee, B.D. New perspectives in studying type 1 diabetes susceptibility biomarkers. *Int. J. Mol. Sci.* **2025**, 26, 3249.

[5] Galicia-Garcia, U.; Benito-Vicente, A.; Jebari, S.; Larrea-Sebal, A.; Siddiqi, H.; Uribe, K.B.; Ostolaza, H.; Martín, C. Pathophysiology of Type 2 Diabetes Mellitus. *Int. J. Mol. Sci.,* **2020**, *21*(17), 6275.
[http://dx.doi.org/10.3390/ijms21176275] [PMID: 32872570]

CHAPTER 5

Disorders of the Kidney and Urinary Tract

Abstract: This chapter offers a comprehensive overview of disorders that affect the kidney and urinary tract. It explores their pathophysiology, clinical manifestations, diagnostic approaches, and treatment options. The chapter emphasizes both acute and chronic conditions, covering common issues such as infections, kidney stones, glomerulonephritis, and chronic kidney disease. It highlights the importance of early detection and intervention to prevent complications and improve patient outcomes. The chapter also discusses advances in diagnostic techniques and therapeutic strategies, as well as preventive measures and patient education. The information aims to equip professionals with the knowledge needed to effectively manage these disorders and improve patient care.

Keywords: Diagnostics, Kidney diseases, Kidney stones, Kidney, Techniques, Urinary tract.

Learning objectives

By the end of this chapter, one should:

1. Be able to describe the principles and control of renal function and the urinary tract, and clinical interventions that supplement/replace declining renal function.
2. Be able to describe the causes, clinical signs, and symptoms of the following features of disorders relating to renal function, the urinary tract, and uric acid metabolism.
3. Be able to describe which laboratory investigations are important in their detection, diagnosis, and management.
4. Be able to recognize the analytical methods available for their measurement.
5. Be able to select laboratory investigations and interpret the analytical results in the context of clinical signs and symptoms.

Simmi Kharb

KEY CONCEPTS

KEY CONCEPTS: Normal Renal Function

Principles and Control of Renal Function (Table 1-3) [1-4]

Table 1. Renal function.

Key Concept	*Details*
Filtration	Glomerular Filtration: Blood is filtered in the glomeruli.
	GFR: Normal rate is 90-120 mL/min.
	Filtrate Composition: Includes water, electrolytes, glucose, amino acids, and waste products.
Reabsorption	Proximal Tubule: Reabsorbs the majority of water, electrolytes, glucose, and amino acids.
	Loop of Henle: Descending limb reabsorbs water; ascending limb reabsorbs sodium and chloride.

Renal function: details.

Table 2. Details of renal function.

Key Concept	*Details*
Secretion	Tubular Secretion: Active transport of substances from blood into tubular fluid.
	Substances Secreted: Hydrogen ions, potassium, ammonia, certain drugs, and toxins.
	Sites: Primarily in the proximal tubule, distal tubule, and collecting duct.
Excretion	Formation of Urine: Final adjustments in the collecting duct; urine flows to the bladder for storage and excretion.
	Urine Composition: Water, urea, creatinine, uric acid, electrolytes, and other metabolic waste products.
Regulation of Blood Pressure	RAAS: Renin released in response to low BP; angiotensin II increases BP; aldosterone increases Na+ reabsorption.
Electrolyte Balance	Sodium (Na+): Regulated by aldosterone.
	Potassium (K+): Regulated by aldosterone.
	Calcium (Ca2+): Reabsorption influenced by PTH.
	Phosphate (PO4^3-): Excretion influenced by PTH.
Acid-Base Balance	Bicarbonate Reabsorption: Occurs in the proximal tubule.
	Hydrogen Ion Secretion: Occurs in the proximal tubule, distal tubule, and collecting duct.
Erythropoiesis	Erythropoietin (EPO) Production: Stimulated by low oxygen levels; promotes red blood cell production.

(Table 2) cont.....

Key Concept	*Details*
Waste Removal	Urea: Byproduct of protein metabolism.
	Creatinine: Byproduct of muscle metabolism.
	Uric Acid: Byproduct of purine metabolism.
Water Balance	Antidiuretic Hormone (ADH): Increases water reabsorption in the collecting ducts; concentrates urine.

Key concepts: Endocrine functions of the kidney.

Table 3. Renal function: endocrine functions.

Endocrine Function	*Details*
Erythropoietin (EPO) Production	Stimulus: Low oxygen levels in the blood.
	Role: Stimulates red blood cell production in the bone marrow.
	Effect: Increases oxygen-carrying capacity of the blood.
Renin-Angiotensin-Aldosterone System (RAAS)	Renin Release: Produced by juxtaglomerular cells in response to low blood pressure or sodium levels.
	Angiotensin II: Formed from angiotensin I, a potent vasoconstrictor, raises blood pressure.
	Aldosterone: Stimulated by angiotensin II, increases sodium and water reabsorption, raises blood pressure.
Vitamin D Activation	Conversion: Converts inactive vitamin D (calcidiol) to the active form (calcitriol).
	Role of Calcitriol: Enhances calcium and phosphate absorption from the gut.
	Effect: Regulates calcium and phosphate homeostasis, essential for bone health.
Prostaglandin Production	Function: Produces prostaglandins, which regulate blood flow within the kidneys.
	Effects: Modulate inflammation, influence kidney function, and affect blood pressure regulation.
Regulation of Blood Pressure	Mechanisms: Through RAAS and the production of prostaglandins.
	Effects: Maintains systemic blood pressure and renal perfusion.
Glucose Homeostasis	Gluconeogenesis: Kidneys produce glucose from non-carbohydrate sources during fasting states.
	Hormonal Influence: Insulin and other hormones regulate this process.

Key concepts: Hemodialysis.

CLINICAL INTERVENTIONS THAT SUPPLEMENT/REPLACE DECLINING RENAL FUNCTION (TABLE 4-8) [1-3]

Table 4. Clinical interventions that supplement/replace declining renal function.

Key Concept	Details
Purpose of Hemodialysis	Definition: Medical procedure using a machine to filter wastes, toxins, and excess fluids from the blood.
	Goal: Maintain chemical balance and prevent waste buildup in patients with kidney failure.

Details:

Table 5. Hemodialysis.

Key Concept	Details
Mechanism of Hemodialysis	Diffusion: Solutes move from a higher concentration (blood) to a lower concentration (dialysate).
	Osmosis: Excess water removed through osmosis *via* a concentration gradient between the blood and dialysate.
Components of the Hemodialysis System	Dialyzer: Contains a semipermeable membrane for waste and fluid exchange.
	Dialysate: A Special fluid that draws out toxins and replenishes necessary chemicals.
	Hemodialysis Machine: Regulates blood and dialysate flow, monitors treatment parameters.
Access for Hemodialysis	AV Fistula: Surgical connection between artery and vein for high blood flow.
	AV Graft: A Synthetic tube connecting an artery to a vein when the veins are unsuitable.
	Central Venous Catheter: Temporary access in a large vein for immediate dialysis needs.
Frequency and Duration	Typical Schedule: Three times a week, each session lasting 3-5 hours.
	Adjustments: Based on the patient's medical condition and needs.
Monitoring and Safety	Vital Signs: Monitoring blood pressure, heart rate, and weight before, during, and after sessions.
	Anticoagulation: Use of heparin or other anticoagulants to prevent blood clotting in the dialysis circuit.

(Table 5) cont.....

Key Concept	Details
Complications and Management	Hypotension: Managed by adjusting fluid removal rates and administering saline.
	Infections: Prevented with aseptic techniques; sometimes requires antibiotic treatment.
	Electrolyte Imbalances: Managed by monitoring and adjusting dialysate composition.
Patient Education and Lifestyle	Diet and Fluid Restrictions: Guidelines to control fluid, potassium, phosphorus, and sodium intake.
	Medication Management: Adjustments needed as some drugs are removed during dialysis.
Impact on Quality of Life	Physical Well-Being: Alleviates symptoms like fatigue, swelling, and difficulty breathing.
	Psychological Support: Counseling and support groups for coping with emotional and social challenges.
Advancements in Hemodialysis	Portable Dialysis Machines: Enable home dialysis or dialysis during travel.
	Improved Membrane Technology: Enhances efficiency and biocompatibility of the dialysis process.

Key concepts: Peritoneal dialysis.

Table 6. Peritoneal dialysis.

Key Concept	Details
Purpose of Peritoneal Dialysis	Definition: Uses peritoneum as a natural semipermeable membrane to filter blood.
	Goal: Maintain fluid, electrolyte, and waste balance in kidney failure patients.
Mechanism of Peritoneal Dialysis	Diffusion: Waste products move from blood to dialysate based on concentration gradients.
	Osmosis: Excess fluid removed from the blood *via* osmotic agents in dialysate.
Types of Peritoneal Dialysis	CAPD: Manual exchanges several times a day.
	APD: Automated exchanges using a cycler, typically overnight.
Components of Peritoneal Dialysis	Dialysate: Sterile solution with glucose or osmotic agents.
	Peritoneal Catheter: A tube placed in the abdominal cavity for dialysate exchange.
	Peritoneal Membrane: Natural lining of the abdomen acting as the filter.

(Table 6) cont.....

Key Concept	Details
Procedure of Peritoneal Dialysis	Filling: Dialysate infused into the peritoneal cavity.
	Dwell Time: Dialysate remains in cavity for waste/fluid exchange.
	Draining: Used dialysate is drained and discarded.
Advantages of Peritoneal Dialysis	Patient Independence: Flexible scheduling can be done at home or work.

Details:

Table 7. Details of peritoneal dialysis.

Key Concept	Details
Advantages of Peritoneal Dialysis	Dietary Freedom: Fewer dietary restrictions.
	Better Hemodynamic Stability: Less impact on blood pressure and the cardiovascular system.
Complications and Management	Peritonitis: Infection prevention with aseptic technique; treated with antibiotics.
	Catheter Issues: Managed by proper care; surgical intervention if necessary.
	Hernias: Managed by surgical repair if needed.
Monitoring and Safety	Infection Control: Aseptic technique to prevent peritonitis.
	Electrolyte Balance: Regular blood chemistry monitoring.
	Volume Status: Monitoring fluid status to avoid dehydration or overload.
Patient Education and Lifestyle	Training: On exchanges and sterile technique.
	Diet/Fluid Management: Guidance to optimize dialysis effectiveness.
	Regular Follow-Up: Routine check-ups to monitor and adjust treatment.
Suitability and Selection	Patient Suitability: Ideal for those with residual kidney function or who prefer home treatment.
	Contraindications: Not suitable for those with extensive abdominal surgery, severe malnutrition, or obesity.

Key concepts: Renal transplant biochemistry.

Table 8. Biochemistry of renal transplant.

Key Concept	Details
Immunology and Histocompatibility	HLA Matching: Reduces rejection risk by matching donor and recipient HLA proteins.
	Crossmatching: Tests for recipient antibodies against donor HLA.
Rejection Mechanisms	Hyperacute Rejection: Minutes to hours, due to pre-existing antibodies.
	Acute Rejection: Days to months, mediated by T cells and B cells.
	Chronic Rejection: Months to years, due to long-term immune response and fibrosis.
Immunosuppressive Therapy	Calcineurin Inhibitors (CIs): Tacrolimus, cyclosporine, inhibit T-cell activation.
	Antiproliferative Agents: Mycophenolate mofetil inhibits DNA synthesis in T and B cells.
	Corticosteroids: Prednisone reduces inflammation and immune response.
	mTOR Inhibitors: Sirolimus inhibits T-cell proliferation.
Biochemical Monitoring	Drug Levels: Ensuring therapeutic range and minimizing toxicity of immunosuppressants.
	Kidney Function Tests: Monitoring creatinine, BUN, and GFR.
	Electrolytes and Metabolites: Monitoring potassium, sodium, calcium, phosphate, and glucose.
Metabolic Complications	Hyperglycemia and Diabetes: From corticosteroids and CIs, managed by glucose monitoring.
	Dyslipidemia: Elevated cholesterol and triglycerides managed by diet, exercise, and medications.
	Hypertension: Common, managed by antihypertensive medications.
Infection Risk	Immunosuppression and Infection: Increased susceptibility; use of prophylactic antibiotics, antivirals, antifungals.
	Cytomegalovirus (CMV): Common post-transplant infection, requiring monitoring and treatment.
Graft Monitoring and Biopsies	Surveillance Biopsies: Detect subclinical rejection and other pathologies.
	Biomarkers: Non-invasive markers for early detection of rejection (*e.g.*, donor-derived cell-free DNA).
Long-Term Graft Health	Fibrosis and Calcineurin Toxicity: Chronic allograft nephropathy from long-term CI use.
	Post-Transplant Malignancies: Increased cancer risk due to long-term immunosuppression.

(Table 8) cont.....

Key Concept	Details
Patient Compliance and Education	Adherence to Medication: Critical for preventing rejection and maintaining graft function.
	Lifestyle Modifications: Diet, exercise, and follow-ups for overall health and transplant success.

BIOCHEMICAL BASIS, RELEVANT LAB TESTS, INTERPRETATION OF RESULTS, AND APPLIED ASPECTS

Laboratory Investigations are Important in their Detection, Diagnosis, and Management (Table 9-10)

Table 9. Laboratory investigations in detection, diagnosis, and management.

Category	Test	Purpose	Details
Detection and Diagnosis	**Serum Creatinine**	Assess kidney function by estimating Glomerular Filtration Rate (GFR).	Elevated levels indicate impaired kidney function. (Figs. **1**; **2**).
	Blood Urea Nitrogen (BUN)	Evaluate kidney function and hydration status.	High levels suggest reduced kidney function or dehydration.
	Estimated Glomerular Filtration Rate (eGFR)	Calculate kidney function based on serum creatinine, age, sex, and race.	Lower values indicate decreased kidney function.
	Urinalysis	Detect abnormalities in urine (*e.g.*, protein, blood, glucose, leukocytes).	Helps identify kidney disease, infection, or other conditions.
	Urine Protein-t--Creatinine Ratio (UPCR)	Quantify protein excretion to assess kidney damage.	A high ratio indicates significant proteinuria, suggestive of kidney damage.
	Urine Albumin-t--Creatinine Ratio (UACR)	Detects microalbuminuria for early kidney damage.	Used to screen for kidney damage, particularly in diabetic patients.
	Electrolytes (Sodium, Potassium, Chloride, Bicarbonate)	Evaluate electrolyte balance and acid-base status.	Imbalances may indicate kidney dysfunction affecting electrolyte regulation.
	Imaging Studies (Ultrasound, CT, MRI)	Visualize kidney structure and detect abnormalities.	Non-invasive methods to detect structural issues or blockages affecting kidney function.
	Renal Biopsy	Histological examination of kidney tissue.	Diagnose specific types of kidney disease and assess severity.

(Table 9) cont.....

Category	Test	Purpose	Details
Management and Monitoring	**Creatinine Clearance Test**	Measure how well the kidneys filter creatinine from the blood.	Provides a more accurate assessment of GFR than serum creatinine alone. (Fig. 3)
	Serum Electrolytes	Monitor levels of electrolytes in the blood.	Regular monitoring is necessary to manage imbalances in patients with kidney disease.
	Serum Calcium and Phosphorus	Assess mineral and bone metabolism in Chronic Kidney Disease (CKD).	Imbalances are common in CKD and can affect bone health.
	Parathyroid Hormone (PTH)	Evaluate secondary hyperparathyroidism in CKD patients.	Elevated PTH indicates disrupted calcium-phosphate metabolism in CKD.
	Vitamin D Levels	Assess vitamin D status, important for bone health in CKD patients.	Deficiency is common in CKD and contributes to bone disease.
	Complete Blood Count (CBC)	Monitor for anemia, which is commonly associated with CKD.	Anemia management is crucial in CKD patients due to reduced erythropoietin production.
	Lipid Profile	Evaluate cardiovascular risk factors in CKD patients.	CKD increases cardiovascular risk; lipid management is important.
	Blood Pressure Monitoring	Regular assessment to manage hypertension in CKD patients.	Hypertension is both a cause and consequence of CKD, requiring strict control.
	24-Hour Urine Collection	Measure total protein, creatinine, and other substances excreted in urine over 24 hours.	Provides comprehensive data on kidney excretory function and proteinuria.
	Dialysis Adequacy Tests (Kt/V, URR)	Assess the effectiveness of dialysis treatment in removing waste products.	Kt/V and URR are used to measure dialysis efficiency and adequacy.

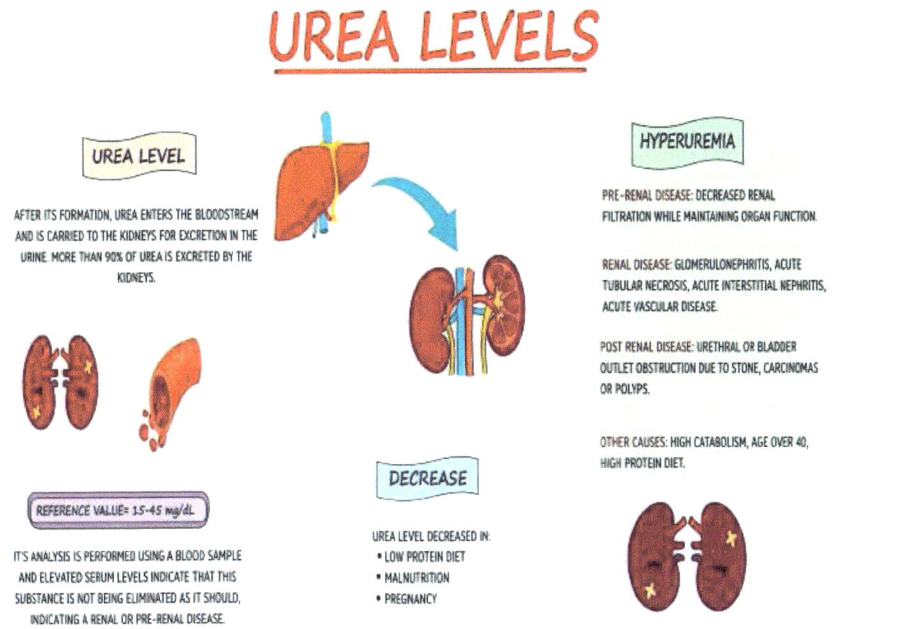

Fig. (1). Urea levels.

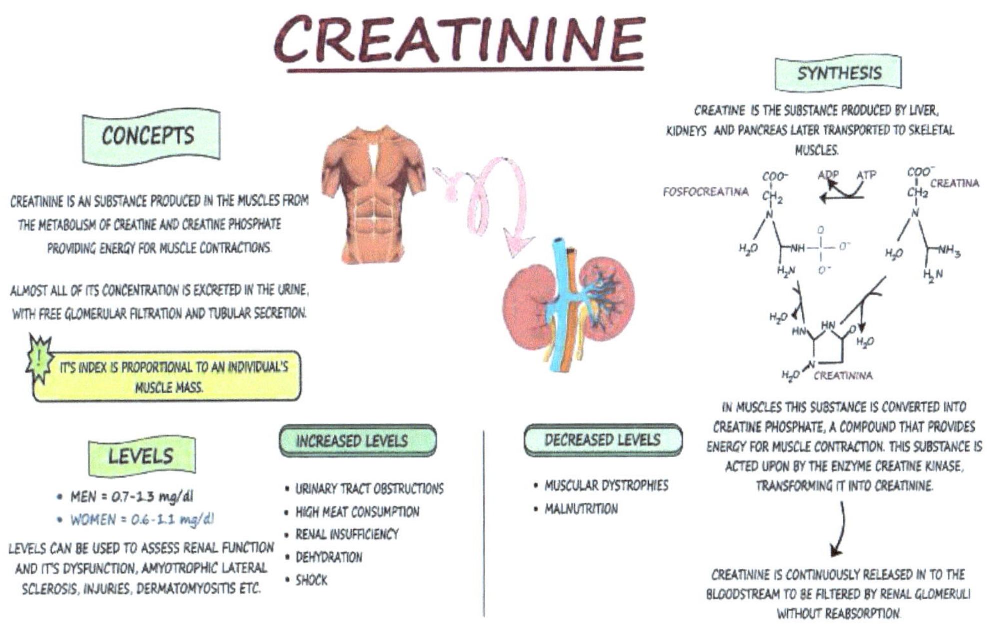

Fig. (2). Creatine levels.

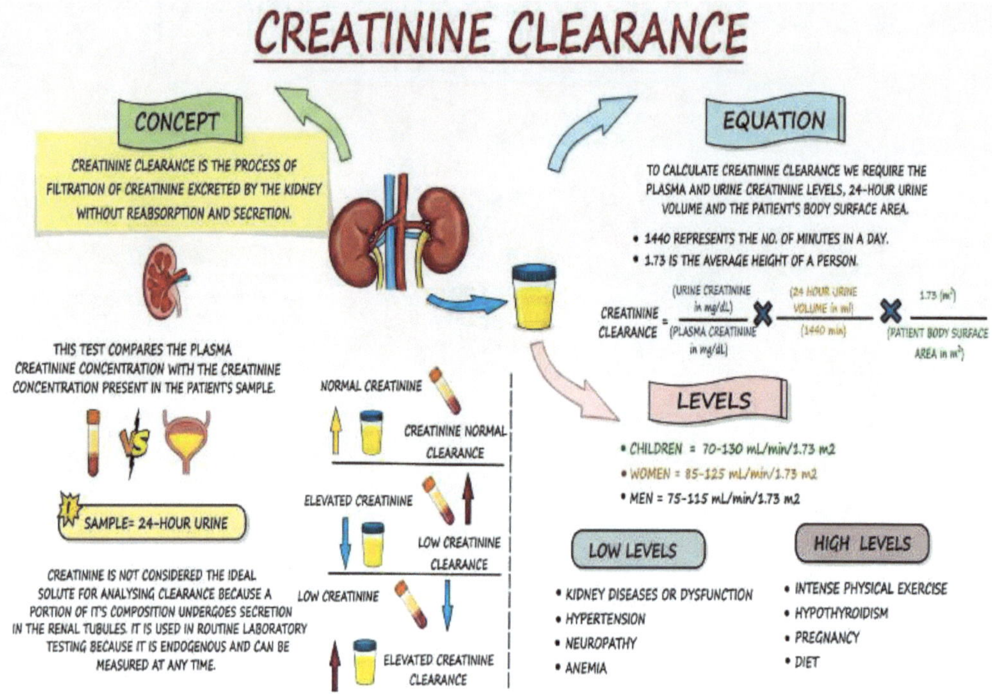

Fig. (3). Creatinine clearance.

Table 10. Specialized laboratory investigations.

Category	Test	Purpose	Details
Specialized Tests	**Cystatin C**	Alternative marker for estimating GFR, less influenced by muscle mass than creatinine.	Useful in patients where creatinine-based estimates may be inaccurate.
	Fibroblast Growth Factor 23 (FGF23)	Marker of mineral metabolism and cardiovascular risk in CKD patients.	Elevated levels are associated with adverse outcomes in CKD.
	Kidney Injury Molecule-1 (KIM-1)	Biomarker for early detection of acute kidney injury.	Useful in identifying early kidney damage before a significant decline in function.
	Neutrophil Gelatinase-Associated Lipocalin (NGAL)	Early biomarker for acute kidney injury.	Increases rapidly in response to kidney injury, providing early detection.

Specialized Tests

ANALYTICAL METHODS

Analytical Methods Available for Measurement (Table 11-13)

Table 11. Analytical methods available for measurement.

Test	Analytical Method
Serum Creatinine	**Spectrophotometry (Jaffe Method, Enzymatic Method)**: Measures creatinine levels based on color change.
	High-Performance Liquid Chromatography (HPLC): Precise quantification of creatinine.
Blood Urea Nitrogen (BUN)	**Spectrophotometry**: Measures urea concentration by colorimetric change after reaction with urease.
Estimated Glomerular Filtration Rate (eGFR)	**Mathematical Estimation**: Calculated using serum creatinine, age, sex, and race equations (*e.g.*, MDRD, CKD-EPI).
Urinalysis	**Dipstick Test**: Rapid screening for multiple analytes (protein, blood, glucose, leukocytes).
	Microscopic Examination: Identifies cells, casts, crystals, and bacteria.
Urine Protein-to-Creatinine Ratio (UPCR)	**Spectrophotometry**: Measures protein and creatinine levels and calculates the ratio.
Urine Albumin-to-Creatinine Ratio (UACR)	**Immunoassay**: Measures albumin levels using antibodies and spectrophotometry for creatinine.
Electrolytes (Sodium, Potassium, Chloride, Bicarbonate)	**Ion-Selective Electrodes (ISE)**: Measure ion concentrations directly in blood samples.
	Flame Photometry: Measures sodium and potassium by flame emission.
Imaging Studies (Ultrasound, CT, MRI)	**Ultrasound**: Uses high-frequency sound waves to visualize kidney structure.
	Computed Tomography (CT): X-ray imaging to create detailed cross-sectional images.
	Magnetic Resonance Imaging (MRI): Uses magnetic fields and radio waves to visualize soft tissues.
Renal Biopsy	**Histopathology**: Microscopic examination of stained kidney tissue sections.
Creatinine Clearance Test	**Spectrophotometry**: Measures creatinine in blood and urine to calculate clearance.
Serum Electrolytes	**Ion-Selective Electrodes (ISE)**: Direct measurement of ion concentrations.

(Table 11) cont.....

Test	Analytical Method
Serum Calcium and Phosphorus	**Spectrophotometry**: Colorimetric assays for calcium and phosphorus levels.
Parathyroid Hormone (PTH)	**Immunoassay (ELISA, Chemiluminescence)**: Measures PTH using specific antibodies.
Vitamin D Levels	**Immunoassay (RIA, ELISA)**: Measures 25-hydroxyvitamin D levels using antibodies.
	Liquid Chromatography-Mass Spectrometry (LC-MS/MS): Highly specific and sensitive measurement.
Complete Blood Count (CBC)	**Automated Hematology Analyzer**: Counts and differentiates blood cells using electrical impedance or flow cytometry.
Lipid Profile	**Enzymatic Colorimetric Assays**: Measures total cholesterol, HDL, LDL, and triglycerides.
Blood Pressure Monitoring	**Sphygmomanometer**: Measures blood pressure manually or automatically.
24-Hour Urine Collection	**Spectrophotometry**: Measures total protein, creatinine, and other substances in collected urine.
Dialysis Adequacy Tests (Kt/V, URR)	**Mathematical Calculation**: Uses pre- and post-dialysis Blood Urea Nitrogen (BUN) levels.
Cystatin C	**Immunoassay (Nephelometry, ELISA)**: Measures cystatin C levels using antibodies.
Fibroblast Growth Factor 23 (FGF23)	**Immunoassay (ELISA, Chemiluminescence)**: Measures FGF23 levels using specific antibodies.
Kidney Injury Molecule-1 (KIM-1)	**Immunoassay (ELISA)**: Measures KIM-1 levels using specific antibodies.
Neutrophil Gelatinase-Associated Lipocalin (NGAL)	**Immunoassay (ELISA)**: Measures NGAL levels using specific antibodies.

Table 12. Laboratory investigations and interpretation of analytical results.

Category	Test	Purpose/ detection	Details
Detection and Diagnosis	**Urinalysis**	Detect abnormalities in urine infection, blood, protein, and glucose	To identify Urinary Tract Infections (UTIs), kidney disease, and other conditions
	Urine Culture	Identify bacterial pathogens causing UTI	To determine specific bacteria and their antibiotic susceptibility
	Microscopic Examination of Urine	Detect cells, casts, crystals, and bacteria in urine	To get detailed information on the type of urinary tract condition

(Table 12) cont.....

Category	Test	Purpose/ detection	Details
	Urine Dipstick Test	Rapid screening for multiple analytes protein, blood, glucose, leukocytes	Quick, initial assessment tool for urinary tract conditions
	Urine Cytology	To detect abnormal cells in urine for possible bladder cancer	For examining urine samples for the presence of cancerous or precancerous cells
	Urine Electrolytes	Assess electrolyte balance and renal tubular function	Measures sodium, potassium, chloride, and other electrolytes in urine
Management and Monitoring	**Urine Protein-to-Creatinine Ratio (UPCR)**	To quantify protein excretion for assessing kidney damage	A high UPCR ratio indicates significant proteinuria, suggestive of kidney damage
	Urine Albumin-t- -Creatinine Ratio (UACR)	Detects microalbuminuria for early kidney damage	For screening kidney damage, especially in diabetics
	Renal Function Tests	For assessing kidney function serum creatinine, eGFR, BUN	For monitoring kidney health and function in patients with urinary tract conditions
	Urine Osmolality	To assess the concentrating ability of the kidneys	High or low osmolality indicates altered kidney function
	Urine pH	To measure the acidity or alkalinity of urine	Abnormal pH is indicative of certain types of kidney stones and infections
	Urine Specific Gravity	To assess urine concentration	Gives information on kidney function and hydration status

Table 13. Analytical methods available.

Test	*Methods*
Urinalysis	**Dipstick Test:** Chemical strips to detect specific substances in urine protein, glucose
	Microscopic Examination Identify cells, casts, crystals, and bacteria
Urine Electrolytes	**Ion-Selective Electrodes (ISE):** Measure ion concentrations directly in urine samples
Urine Protein-to-Creatinine Ratio (UPCR)	**Spectrophotometry:** Measure protein and creatinine levels, calculate the ratio

(Table 13) cont.....

Test	*Methods*
Urine Albumin-to-Creatinine Ratio (UACR)	**Immunoassay:** Measure albumin levels using antibodies **Spectrophotometry:** Creatinine
Renal Function Tests	**Spectrophotometry:** Measure serum creatinine, BUN, and other markers of kidney function
	Mathematical Estimation: eGFR calculation using serum creatinine, age, sex, and race equations
Urine Osmolality	**Osmometry:** Measures the concentration of solutes in urine by freezing point depression
Urine pH	**pH Meter:** Measure the hydrogen ion concentration for acidity or alkalinity
Urine Specific Gravity	**Refractometry:** Measure the concentration of solutes in urine by light refraction

BIOCHEMICAL BASIS, RELEVANT LAB TESTS, INTERPRETATION, AND APPLIED ASPECTS: URINARY TRACT

Laboratory Investigations and Interpret the Analytical Results in the Context of Clinical Signs and Symptoms (Table 12-15, 16) [1-4]

Table 14. Analytical methods available.

Category	**Test**	**Purpose**	**Details**
Detection and Diagnosis	**Serum Uric Acid (SUA)**	To measure uric acid levels in the blood	Elevated levels may indicate gout, kidney disease, or other metabolic disorders
	Urine Uric Acid	To quantify uric acid excretion in urine	Provides insight into the handling of uric acid and kidney function.
	Urinalysis	To detect uric acid crystals/ urate crystals in urine	Aids in the diagnosis of gout and monitoring of treatment
	Renal Function Tests	To assess kidney function in relation to uric acid metabolism: serum creatinine, BUN	Evaluates kidney health and its impact on uric acid levels
	Synovial Fluid Analysis	To detect uric acid crystals in synovial fluid	For diagnosing gout by confirming the presence of urate crystals in joints

(Table 14) cont.....

Category	Test	Purpose	Details
Management and Monitoring	**Urine pH**	To measure the acidity or alkalinity of urine affecting uric acid solubility	To manage uric acid crystallization and deposition
	Serum Electrolytes	To monitor electrolyte imbalances affecting uric acid excretion: potassium, calcium	To ensure balanced conditions for uric acid metabolism
	Liver Function Tests	To assess the influence of the liver on uric acid production and metabolism	Elevated enzymes (AST, ALT) may suggest liver dysfunction affecting uric acid levels
Analytical Methods	**Enzymatic Colorimetric Assay**	Quantifies uric acid levels Based on the color change reaction	For accurate measurement of serum or urine uric acid levels
	High-Performance Liquid Chromatography (HPLC)	Precise quantification of uric acid levels	To measure uric acid with high sensitivity and specificity
	Spectrophotometry	To measure the absorbance of uric acid at specific wavelengths	To determine the uric acid concentration in biological samples
	Ion-Selective Electrodes (ISE)	To measure ion concentrations in urine or serum, including uric acid	Direct method to assess uric acid levels, especially in a urine sample
	Mass Spectrometry (MS)	To identify and quantify uric acid and its metabolites with high accuracy	Useful for research purposes and in specialized clinical settings requiring precise measurements
	Fluorescence Detection	To detect uric acid with high sensitivity in certain assays	For enhanced detection capabilities in specific laboratory setups

Table 15. Analytical methods available.

Disorder	Laboratory Tests	Interpretation of Analytical Results
Acute Kidney Injury (AKI)	Serum creatinine, urine output monitoring, and urinalysis	Elevated creatinine and decreased urine output indicate *impaired* kidney function
Chronic Kidney Disease (CKD)	Serum creatinine, eGFR calculation, urinalysis, and urine albumin	Decreased eGFR and the presence of albuminuria suggest *kidney damage and progression of CKD*
Amyloidosis	Congo red staining of biopsy samples	The presence of *apple-green birefringence* under polarized light confirms amyloid deposition

(Table 15) cont.....

Disorder	Laboratory Tests	Interpretation of Analytical Results
ANCA-Associated Vasculitis	ANCA (Anti-Neutrophil Cytoplasmic Antibodies)	Positive ANCA titres indicate *autoimmune vasculitis* affecting small blood vessels
Cryoglobulin-Associated Renal Damage	Cryoglobulins in serum, complement C3 and C4	Detects cryoglobulins and assesses complement levels; associated with renal damage in cryoglobulinemia
Drug-Induced Renal Damage	Measurement of drug levels, *e.g.*, NSAIDs, ACE inhibitors	Elevated drug levels or changes in renal function markers suggest drug-induced nephrotoxicity
Glomerular Dysfunction/Glomerulonephritis	Urinalysis, proteinuria, serum complement levels	The presence of proteinuria and abnormalities in complement levels indicates glomerular damage or inflammation
Goodpasture's Syndrome	Anti-Glomerular Basement Membrane (GBM) antibodies	Positive anti-GBM antibodies confirm an autoimmune attack on the kidney's basement membrane
Gout/Hyperuricemia	Serum uric acid, joint fluid analysis (for urate crystals)	Elevated uric acid levels and the presence of urate crystals in joints confirm gout or hyperuricemia
Myeloma-Associated Renal Damage	Serum and urine protein electrophoresis, serum free light chains	Detects abnormal protein bands or free light chains associated with myeloma-related kidney damage
Proteinuria/Glomerular Permeability	Urine protein, urine albumin, urine dipstick analysis	Detects excess protein in urine, indicating glomerular damage or increased permeability
Nephritic Syndrome	Urinalysis, anti-Streptolysin O (ASO) titre	Hematuria, proteinuria, and elevated ASO titers indicate acute glomerulonephritis
Nephrotic Syndrome	Urinalysis, urine albumin, serum albumin, lipid profile	Severe proteinuria (>3.5 g/day), hypoalbuminemia, and lipid abnormalities characterize nephrotic syndrome
Diabetic Nephropathy	Urine Albumin-to-Creatinine ratio (UACR), serum creatinine	Elevated UACR and progressive decline in eGFR indicate diabetic kidney damage
Renal Tubular Disease	Urine electrolytes (*e.g.*, phosphate, calcium, magnesium), urine pH	Electrolyte imbalances and abnormal urine pH suggest renal tubular dysfunction
Diseases of Renal Endocrine Dysfunction	Erythropoietin levels, renin levels	Abnormal levels indicate dysfunction in the kidney's endocrine functions affecting erythropoiesis or BP control

(Table 15) cont.....

Disorder	Laboratory Tests	Interpretation of Analytical Results
Systemic lupus erythematosus (SLE)	Anti-nuclear antibodies (ANA), anti-dsDNA antibodies	Positive ANA and anti-dsDNA antibodies suggest lupus nephritis affecting the kidneys
Urinary Tract Infection (UTI)	Urinalysis, urine culture	The presence of bacteria, leukocytes, and symptoms indicates a urinary tract infection
Renal Stone Formation	Urinalysis, stone analysis (composition)	The presence of crystals or stones in urine, along with their composition, guides management strategies.
Uremia	Blood Urea Nitrogen (BUN), serum creatinine, electrolytes	Elevated BUN and creatinine with electrolyte imbalances indicate uremic symptoms
Vasculitis (ANCA-associated)	ANCA, complement C3 and C4 Levels	Positive ANCA and abnormal complement levels suggest ANCA-associated vasculitis affecting kidneys

ANALYTICAL METHODS FOR URINARY TRACT TESTS

Relevant Laboratory Tests for Uric Acid Metabolism

Relevant Laboratory Tests for Detection, Diagnosis, Management, and Monitoring: Renal and Related Disorders

Summary

This chapter covers the etiology, pathophysiology, clinical manifestations, and management of common kidney and urinary tract disorders, such as urinary tract infections, kidney stones, chronic kidney disease, and acute renal failure. Diagnostic approaches, treatment modalities, and the impact of these disorders on overall health and quality of life are discussed.

Exercises

1. What are the potential causes of elevated creatinine levels?
2. Discuss the clinical significance of red blood cell casts in the urine.
3. Describe the pathophysiological process leading from acute glomerulonephritis to chronic kidney disease.
4. What key biochemical markers are used to diagnose acute kidney injury?
5. How is the Glomerular Filtration Rate (GFR) calculated, and what are the limitations of using creatinine as a marker?

6. Explain the clinical significance of the urinary Albumin-to-Creatinine Ratio (ACR) in diagnosing chronic kidney disease.
7. Describe the biochemical differences between nephrotic and nephritic syndromes.
8. What are the common causes of metabolic acidosis in patients with renal failure?

CONCLUSION

In conclusion, understanding kidney and urinary tract disorders is crucial for effective management and prevention. Early diagnosis and appropriate treatment can significantly improve patient outcomes. Emphasising patient education and lifestyle changes is essential for preventing disease progression and enhancing quality of life.

REFERENCES

[1] Brenner & Rector's The Kidney Volume I and Volume II. 12th edition. Edited by Barry M. Brenner. Elsevier; **2026**.

[2] Rose B.D., Post T.W. Clinical Physiology of Acid-Base and Electrolyte Disorders. 6th edition, Blackwell's; **2026**.

[3] Pham P.C.T., Pham P.T.T. Nephrology and Hypertension Board Review. 3rd edition, Wolters Kluwer; **2025**.

[4] Lerma E.V., Sparks M.A., Topf J.M. Nephrology Secrets. 5th Edition, Elsevier; **2026**.

CHAPTER 6

Endocrinology

Abstract: Endocrinology, a critical branch of medicine, deals with the endocrine system, its diseases, and the secretions known as hormones. It plays a vital role in maintaining homeostasis. This chapter delves into the practical aspects of endocrinology, specifically focusing on prevalent disorders such as diabetes mellitus, thyroid diseases, adrenal insufficiency, and growth disorders. The emphasis is placed on the biochemical pathways, diagnostic approaches, and therapeutic strategies. Significant advances in molecular diagnostics and personalized medicine have greatly enhanced the understanding and treatment of endocrine disorders, ultimately leading to improved patient outcomes.

Keywords: Adrenal insufficiency, Diseases, Diabetes mellitus, Endocrinology, Growth disorders, Lab diagnosis, Thyroid diseases.

LEARNING OBJECTIVES

By the end of this chapter, one should be able to:

- Understand and describe all aspects of hormone action and regulation.
- Understand the causes, clinical presentation, and laboratory diagnosis of endocrine disorders.
- Understand the relevance of laboratory investigations in the detection, diagnosis, and management of each endocrine disorder.
- Recognize the analytical methods available for their measurement.
- Select appropriate laboratory investigations and interpret the analytical results in the context of the clinical signs and symptoms in the patient.

KEY CONCEPTS: HORMONE ACTION

Hormone Action and Regulation (Table 1,2) [1 - 3]

Simmi Kharb

Table 1. Hormone action and regulation.

Key Concept	*Feedback Inhibition*	*Other Regulatory Mechanisms*
Hypothalamic-Pituitary Axis	Hormone levels inhibit hypothalamic/pituitary release	Hypothalamic hormones, circadian rhythms, and neural inputs
Anterior Pituitary Hormones	Target gland hormones provide negative feedback	Releasing hormones (*e.g.*, TRH, CRH, GnRH) and inhibitory hormones (*e.g.*, somatostatin, dopamine)
Pituitary-Hypothalamus Communication	Feedback from target hormones	Anterior: *via* the hypophyseal portal system; Posterior: neural control
Hirsutism and Virilization	Androgens regulate LH/FSH levels	Abnormal feedback loops and direct gland dysfunction
Hypothalamic Hormones	Target gland hormones inhibit hypothalamic hormones	Neural inputs, stress, and circadian rhythms
Inhibitory Hormones	Target gland hormones inhibit hypothalamic/pituitary hormones	Somatostatin and dopamine inhibit GH and prolactin
Posterior Pituitary Hormones	Feedback from osmoreceptors, uterine stretch receptors	Neural signals from the hypothalamus control the release
Primary *vs.* Secondary Causes	Different patterns of hormone levels in primary *vs.* secondary disorders	Target gland hormone levels affect upstream regulation differently
Renin-Angiotensin-Aldosterone Pathway	Elevated BP/sodium inhibits renin release, reducing aldosterone	Blood pressure, sodium, and potassium levels regulate renin; aldosterone adjusts sodium/water reabsorption
Steroid Biosynthesis Pathway	Cortisol provides negative feedback to inhibit ACTH and CRH	Enzymatic control within the adrenal cortex and feedback from end products
Stimulation Tests	Assess gland responsiveness to feedback dysregulation	Administer stimulating agents (*e.g.*, ACTH, TRH) to test functionality
Suppression Tests	Evaluate the ability of feedback to suppress hormone release	Administer suppressive agents (*e.g.*, dexamethasone) to test negative feedback mechanisms

KEY CONCEPTS: HYPOTHALAMIC- PITUITARY AXIS

Table 2. Key concepts of hormone action and regulation.

Key Concept	*Feedback Inhibition*	*Other Regulatory Mechanisms*
Hypothalamic-Pituitary Axis	Hormone levels inhibit hypothalamic/pituitary release	Hypothalamic hormones, circadian rhythms, and neural inputs
Anterior Pituitary Hormones	Target gland hormones provide negative feedback	Releasing hormones (*e.g.*, TRH, CRH, GnRH), inhibitory hormones (*e.g.*, somatostatin, dopamine)
Communication Between Pituitary and Hypothalamus – Anterior *vs.* Posterior	Feedback from target hormones	Anterior: *via* the hypophyseal portal system; Posterior: neural control
Hirsutism and Virilization	Androgens regulate LH/FSH levels	Abnormal feedback loops, direct gland dysfunction
Hypothalamic Hormones	Target gland hormones inhibit hypothalamic hormones	Neural inputs, stress, and circadian rhythms
Inhibitory Hormones	Target gland hormones inhibit hypothalamic/pituitary hormones	Somatostatin and dopamine inhibit GH and prolactin
Posterior Pituitary Hormones	Feedback from osmoreceptors and uterine stretch receptors	Neural signals from the hypothalamus control the release
Primary *vs.* Secondary Causes	Different patterns of hormone levels in primary *vs.* secondary disorders	Target gland hormone levels affect upstream regulation differently
Renin-Angiotensin-Aldosterone Pathway	Elevated BP/sodium inhibits renin release, reducing aldosterone	Blood pressure, sodium, and potassium levels regulate renin; aldosterone adjusts sodium/water reabsorption
Steroid Biosynthesis Pathway	Cortisol provides negative feedback to inhibit ACTH and CRH	Enzymatic control within the adrenal cortex, feedback from end products
Stimulation Tests	Assess gland responsiveness to feedback dysregulation	Administer stimulating agents (*e.g.*, ACTH, TRH) to test functionality
Suppression Tests	Evaluate the ability of feedback to suppress hormone release	Administer suppressive agents (*e.g.*, dexamethasone) to test negative feedback mechanisms

BIOCHEMICAL BASIS, RELEVANT LAB TESTS, INTERPRETATION, AND APPLIED ASPECTS: ENDOCRINE DISORDERS

Causes, Clinical Presentation, and Laboratory Diagnosis of Endocrine Disorders (Table 3). [1 - 4]

Table 3. Causes, clinical presentation, and laboratory diagnosis of endocrine disorders.

Disease/Disorder	Biochemical Changes	Causes	Clinical Signs and Symptoms
Acromegaly/Gigantism	Elevated GH and IGF-1 levels	Pituitary adenoma	Enlarged hands/feet, facial features, joint pain, and organ enlargement
Addison's Syndrome	Low cortisol and aldosterone levels and high ACTH	Autoimmune destruction, infections, and adrenal hemorrhage	Fatigue, weight loss, hypotension, hyperpigmentation, and salt craving
Adrenal Insufficiency	Low cortisol levels, possibly low aldosterone, and androgens	Addison's disease, secondary due to pituitary dysfunction	Fatigue, weakness, hypotension, anorexia, nausea, and hyperpigmentation
Congenital Adrenal Hyperplasia	Elevated 17-OH progesterone, low cortisol, and high androgens	21-hydroxylase deficiency, a genetic mutation	Ambiguous genitalia in females, early virilization in males, and salt wasting
Cushing's Syndrome and Disease	Elevated cortisol, low ACTH (syndrome), and high ACTH (disease)	Exogenous steroids (syndrome) and pituitary adenoma (disease)	Weight gain, moon face, buffalo hump, hypertension, and glucose intolerance
Diabetes Insipidus	Low ADH levels or renal insensitivity to ADH	Central DI (pituitary damage) and nephrogenic DI (renal cause)	Polyuria, polydipsia, dehydration, and dilute urine
Female Infertility	Varied depending on cause (*e.g.*, low FSH/LH and elevated prolactin)	PCOS, hypothalamic dysfunction, and hyperprolactinemia	Irregular menstrual cycles, anovulation, hirsutism, and galactorrhea
Growth Hormone Deficiency	Low GH and IGF-1 levels	Pituitary dysfunction and genetic mutations	Short stature in children, increased fat, and decreased muscle mass
Hyperaldosteronism	Elevated aldosterone and low renin	Adrenal adenoma and bilateral adrenal hyperplasia	Hypertension, hypokalemia, muscle weakness, and fatigue
Hyperprolactinemia	Elevated prolactin levels	Pituitary adenoma, hypothyroidism, and medications	Galactorrhea, amenorrhea, infertility, and decreased libido

(Table 3) cont.....

Disease/Disorder	Biochemical Changes	Causes	Clinical Signs and Symptoms
Hyperthyroidism	Elevated T3/T4 and low TSH	Graves' disease, toxic adenoma, and thyroiditis	Weight loss, heat intolerance, palpitations, tremors, and exophthalmos
Hypothyroidism	Low T3/T4 and high TSH	Hashimoto's thyroiditis, iodine deficiency, and thyroidectomy	Fatigue, weight gain, cold intolerance, constipation, and dry skin
Non-thyroidal Illness	Low T3, normal/low T4, and normal/low TSH	Severe illness and stress	Symptoms of underlying illness and altered thyroid function tests
Male Infertility	Varied depending on cause (*e.g.*, low testosterone, high FSH/LH)	Varicocele, genetic factors, and hormonal imbalances	Low sperm count, poor sperm motility, and erectile dysfunction
Pheochromocytoma	Elevated catecholamines (epinephrine, norepinephrine)	Adrenal medulla tumor	Hypertension, palpitations, sweating, and headaches
Polycystic Ovarian Syndrome	Elevated androgens, LH imbalance, and insulin resistance	Genetic predisposition and metabolic factors	Irregular periods, hirsutism, acne, obesity, and infertility
Premature Ovarian Failure	Low estrogen and high FSH	Autoimmune, genetic factors, and chemotherapy	Amenorrhea, hot flashes, infertility, and vaginal dryness
Sheehan's Syndrome	Low pituitary hormones (*e.g.*, ACTH, TSH, GH)	Postpartum pituitary necrosis	Fatigue, inability to lactate, amenorrhea, and hypotension
SIADH	Low serum sodium, low serum osmolality, and high urine osmolality	CNS disorders, lung disease, and medications	Hyponatremia, confusion, seizures, and decreased urine output
Thyroid Cancer	Elevated calcitonin (medullary type) and thyroid nodule	Genetic mutations and radiation exposure	Thyroid nodule, neck swelling, dysphagia, and hoarseness

RELEVANT LABORATORY TESTS FOR DETECTION, DIAGNOSIS, MANAGEMENT, AND MONITORING

Relevance of Laboratory Investigations in Detection, Diagnosis, and Management of Endocrine Disorders (Table 4). [1 - 3]

Table 4. Relevance of laboratory investigations.

Disease/Disorder	*Biochemical Changes*	*Causes*	*Clinical Signs and Symptoms*
Acromegaly/Gigantism	Elevated GH and IGF-1 levels	Pituitary adenoma	Enlarged hands/feet, facial features, joint pain, and organ enlargement
Addison's Syndrome	Low cortisol and aldosterone levels and high ACTH	Autoimmune destruction, infections, and adrenal hemorrhage	Fatigue, weight loss, hypotension, hyperpigmentation, and salt craving
Adrenal Insufficiency	Low cortisol levels, possibly low aldosterone, and androgens	Addison's disease, secondary due to pituitary dysfunction	Fatigue, weakness, hypotension, anorexia, nausea, and hyperpigmentation
Congenital Adrenal Hyperplasia	Elevated 17-OH progesterone, low cortisol, and high androgens	21-hydroxylase deficiency and genetic mutation	Ambiguous genitalia in females, early virilization in males, and salt wasting
Cushing's Syndrome and Disease	Elevated cortisol, low ACTH (syndrome), and high ACTH (disease)	Exogenous steroids (syndrome) and pituitary adenoma (disease)	Weight gain, moon face, buffalo hump, hypertension, and glucose intolerance
Diabetes Insipidus	Low ADH levels or renal insensitivity to ADH	Central DI (pituitary damage) and nephrogenic DI (renal cause)	Polyuria, polydipsia, dehydration, and dilute urine
Female Infertility	Varied depending on cause (*e.g.*, low FSH/LH and elevated prolactin)	PCOS, hypothalamic dysfunction, and hyperprolactinemia	Irregular menstrual cycles, anovulation, hirsutism, and galactorrhea
Growth Hormone Deficiency	Low GH and IGF-1 levels	Pituitary dysfunction and genetic mutations	Short stature in children, increased fat, and decreased muscle mass
Hyperaldosteronism	Elevated aldosterone and low renin	Adrenal adenoma and bilateral adrenal hyperplasia	Hypertension, hypokalemia, muscle weakness, and fatigue
Hyperprolactinemia	Elevated prolactin levels	Pituitary adenoma, hypothyroidism, and medications	Galactorrhea, amenorrhea, infertility, and decreased libido

(Table 4) cont.....

Disease/Disorder	*Biochemical Changes*	*Causes*	*Clinical Signs and Symptoms*
Hyperthyroidism	Elevated T3/T4 and low TSH	Graves' disease, toxic adenoma, and thyroiditis	Weight loss, heat intolerance, palpitations, tremors, and exophthalmos
Hypothyroidism	Low T3/T4 and high TSH	Hashimoto's thyroiditis, iodine deficiency, and thyroidectomy	Fatigue, weight gain, cold intolerance, constipation, and dry skin
Non-thyroidal Illness	Low T3, normal/low T4, and normal/low TSH	Severe illness and stress	Symptoms of underlying illness and altered thyroid function tests
Male Infertility	Varied depending on cause (*e.g.*, low testosterone, high FSH/LH)	Varicocele, genetic factors, and hormonal imbalances	Low sperm count, poor sperm motility, and erectile dysfunction
Pheochromocytoma	Elevated catecholamines (epinephrine, norepinephrine)	Adrenal medulla tumor	Hypertension, palpitations, sweating, and headaches
Polycystic Ovarian Syndrome	Elevated androgens, LH imbalance, and insulin resistance	Genetic predisposition and metabolic factors	Irregular periods, hirsutism, acne, obesity, and infertility
Premature Ovarian Failure	Low estrogen and high FSH	Autoimmune, genetic factors, and chemotherapy	Amenorrhea, hot flashes, infertility, and vaginal dryness
Sheehan's Syndrome	Low pituitary hormones (*e.g.*, ACTH, TSH, and GH)	Postpartum pituitary necrosis	Fatigue, inability to lactate, amenorrhea, and hypotension
SIADH	Low serum sodium, low serum osmolality, and high urine osmolality	CNS disorders, lung disease, and medications	Hyponatremia, confusion, seizures, and decreased urine output
Thyroid Cancer	Elevated calcitonin (medullary type) and thyroid nodule	Genetic mutations and radiation exposure	Thyroid nodule, neck swelling, dysphagia, and hoarseness

RELEVANT LABORATORY TESTS, INTERPRETATION OF ANALYTICAL RESULTS, AND ANALYTICAL METHODS

Laboratory Investigations and Interpretation of Analytical Results in the Context of the Clinical Signs and Symptoms (Table 5). [1 - 5]

Table 5. Relevance of laboratory investigations, interpretation, and analytical methods available.

Test	*Relevant Laboratory Tests*	*Interpretation of Analytical Results*	*Analytical Methods*
17-Hydroxyprogesterone	Measurement of 17-OHP levels	Elevated in Congenital Adrenal Hyperplasia (CAH)	Immunoassay and LC-MS/MS
Adrenocorticotropic hormone (ACTH)	Measurement of ACTH levels	Elevated in Cushing's disease and low in adrenal insufficiency	Immunoassay, CLIA, and RIA
Anti-diuretic hormone (ADH) / Arginine vasopressin	Measurement of ADH levels	Low in diabetes insipidus and high in SIADH	Immunoassay and RIA
Aldosterone, Renin, and Ratio	Measurement of aldosterone and renin levels and calculation of ratio	High aldosterone/renin ratio in primary hyperaldosteronism	Immunoassay and LC-MS/MS
Androstenedione	Measurement of androstenedione levels	Elevated in adrenal hyperplasia and PCOS	Immunoassay and LC-MS/MS
Angiotensin Converting Enzyme (ACE)	Measurement of ACE levels	Elevated in sarcoidosis and low in ACE inhibitor therapy	Spectrophotometric assay and ELISA
Autoantibodies to the Relevant Endocrine Organ	Detection of specific autoantibodies (*e.g.*, TPO, TG, insulin autoantibodies)	Presence indicates autoimmune thyroiditis, Type 1 diabetes, *etc.*	ELISA, CLIA, and IFA
Catecholamines (Plasma and Urine)	Measurement of plasma and urinary catecholamines	Elevated in pheochromocytoma	HPLC, LC-MS/MS and Immunoassay
Cortisol – Serum, Urine, Salivary	Measurement of cortisol levels in serum, 24-hour urine, and salivary samples	Elevated in Cushing's syndrome, low in adrenal insufficiency	Immunoassay and LC-MS/MS
Dexamethasone Suppression Test	Measurement of cortisol levels after dexamethasone administration	Lack of suppression in Cushing's syndrome	Immunoassay and LC-MS/MS

(Table 5) cont.....

Test	*Relevant Laboratory Tests*	*Interpretation of Analytical Results*	*Analytical Methods*
Dehydroepiandrosterone sulfate (DHEAS)	Measurement of DHEAS levels	Elevated in adrenal hyperplasia, PCOS	Immunoassay and LC-MS/MS
Estradiol	Measurement of estradiol levels	Low in ovarian failure and high in estrogen-producing tumors	Immunoassay and LC-MS/MS
Follicle Stimulating Hormone (FSH)	Measurement of FSH levels	Elevated in primary gonadal failure and low in secondary hypogonadism	Immunoassay, CLIA, and RIA
Growth Hormone (GH)	Measurement of GH levels	Elevated in acromegaly/gigantism and low in GH deficiency	Immunoassay, CLIA, and RIA
Growth Hormone Suppression Test (OGTT)	Measurement of GH levels after oral glucose load	Lack of suppression in acromegaly	Immunoassay, CLIA, and RIA
Insulin-like Growth Factor 1 (IGF-1)	Measurement of IGF-1 levels	Elevated in acromegaly and low in GH deficiency	Immunoassay and LC-MS/MS
Luteinizing Hormone (LH)	Measurement of LH levels	Elevated in primary gonadal failure and low in secondary hypogonadism	Immunoassay, CLIA, and RIA
Metanephrines (Plasma and Urine)	Measurement of plasma and urinary metanephrines	Elevated in pheochromocytoma	HPLC, LC-MS/MS, and Immunoassay
Progesterone	Measurement of progesterone levels	Elevated in the luteal phase and pregnancy	Immunoassay and LC-MS/MS
Prolactin and Macroprolactin	Measurement of prolactin levels and detection of macroprolactin	Elevated in prolactinoma and macroprolactinemia	Immunoassay, PEG precipitation, and LC-MS/MS
Sex Hormone Binding Globulin (SHBG)	Measurement of SHBG levels	Elevated in hyperthyroidism and low in insulin resistance	Immunoassay and ELISA
Testosterone – Total, Free, Bioavailable	Measurement of total, free, and bioavailable testosterone levels	Elevated in PCOS and low in hypogonadism	Immunoassay and LC-MS/MS
Thyroglobulin (TG), Anti-Thyroid Stimulating Hormone (TSH), Free T3, Free T4, Total T3, Total T4	Measurement of thyroid hormones and antibodies	Varied depending on thyroid status (hyperthyroidism and hypothyroidism)	Immunoassay, LC-MS/MS, and ELISA
TSH Receptor Antibodies	Detection of TSH receptor antibodies	Elevated in Graves' disease	Immunoassay and ELISA

(Table 5) cont.....

Test	*Relevant Laboratory Tests*	*Interpretation of Analytical Results*	*Analytical Methods*
Thyroperoxidase (TPO) Antibodies	Detection of TPO antibodies	Elevated in Hashimoto's thyroiditis	Immunoassay and ELISA

SUMMARY

Endocrinology is the comprehensive study of glands and hormones that regulate various bodily functions. This field meticulously addresses disorders such as diabetes, thyroid abnormalities, and adrenal insufficiencies, which are diagnosed through meticulous biochemical tests and advanced imaging studies. The management of these disorders involves decisive hormone replacement therapies, precise surgical interventions, and necessary lifestyle modifications. Recent advancements in genetics and molecular biology have introduced highly targeted therapies and significantly improved diagnostic accuracy. Understanding the integration of hormonal regulation and its clinical implications is crucial for delivering effective treatment and optimal patient care in endocrinology.

EXERCISES

- The roles of insulin and the pathophysiology of diabetes mellitus.
- Biochemical tests for diagnosing thyroid disorders.
- Clinical features and laboratory findings in Addison's disease.
- Diagnosis and treatment of growth hormone deficiency in children.
- Role of molecular diagnostics in managing endocrine cancers.
- Interpretation of laboratory results in hyperparathyroidism.
- How genetic mutations affect endocrine function and lead to disease.
- Therapeutic approaches for managing hyperthyroidism.
- Implications of adrenal insufficiency on electrolyte balance.
- Impact of lifestyle modifications on managing metabolic syndrome.

CONCLUSION

The field of endocrinology is essential for understanding and managing disorders related to hormonal imbalances. Advances in diagnostic techniques and personalized medicine have revolutionized the approach to endocrine diseases. Continuous research and development are crucial for further advancements in this field, aiming to enhance patient care and treatment outcomes. Comprehensive knowledge of endocrine function and its disorders is essential for clinical practice in laboratory medicine.

REFERENCES

[1] Melmed, S.; Polonsky, K.S.; Larsen, P.R.; Kronenberg, H.M. *Williams Textbook of Endocrinology*; 15th edition; **2024**.

[2] Jameson, J.L.; De Groot, L. *J. Endocrinology: Adult and Pediatric,* 7th ed; Saunders, **2016**.

[3] Gardner, D.G.; Shoback, D. *Greenspan's Basic and Clinical Endocrinology,* 10th ed; McGraw-Hill Education, **2020**.

[4] *Fundamentals of Clinical Chemistry and Molecular Diagnostics,* 9th ed; Elsevier, **2026**.

[5] Holt, R.I.G.; Hanley, N.A. *Essential Endocrinology and Diabetes,* 7th ed; Wiley-Blackwell, **2022**.

CHAPTER 7

Fluid and Electrolyte Disorders

Abstract: Understanding and managing fluid and electrolyte disorders is crucial in clinical practice. This chapter explores the pathophysiology, diagnostic approaches, and management strategies for common disorders such as hyponatremia, hyperkalemia, and dehydration. Integration of biochemical data with clinical presentations is emphasized for comprehensive care planning.

Keywords: Fluid, Electrolyte, Hyponatremia, Hyperkalaemia, Dehydration.

- Understand the physiological mechanisms regulating fluid and electrolyte balance.
- Identify the causes and consequences of fluid and electrolyte disorders.
- Describe the laboratory tests used to diagnose fluid and electrolyte imbalances.
- Develop treatment strategies for managing fluid and electrolyte disorders.
- Apply clinical knowledge to interpret laboratory results in the context of patient symptoms.
- Evaluate the impact of fluid and electrolyte disorders on overall health and disease outcomes.

KEY CONCEPTS

Basic Concept and Homeostasis Of Fluid And Electrolyte Balance

Physiological Mechanisms Regulating Fluid And Electrolyte Balance (Table 1) [1-3]

Simmi Kharb

Table 1. Physiological mechanisms regulating fluid and electrolyte balance.

Key Concept	Description	Mechanisms and Regulation
Total Body Water	Comprises about 60% of body weight, divided into Intracellular Fluid (ICF) and Extracellular Fluid (ECF).	Regulated by intake and loss of fluids *via* the kidneys, skin, lungs, and GI tract.
Intracellular Fluid (ICF)	Fluid within cells, makes up about 40% of body weight.	Maintained by cell membrane permeability and osmotic gradients.
Extracellular Fluid (ECF)	Fluid outside cells, includes interstitial fluid, plasma, and transcellular fluid, makes up about 20% of body weight.	Regulated by capillary filtration and reabsorption, lymphatic drainage.
Osmolality	Measure of solute concentration in body fluids, normally around 275-295 mOsm/kg.	Controlled by ADH, the thirst mechanism, and renal function.
Sodium (Na^+)	Major cation in ECF, crucial for maintaining fluid balance, nerve function, and muscle contraction.	Regulated by aldosterone, natriuretic peptides, and ADH.
Potassium (K^+)	A major cation in ICF, essential for cell function, nerve impulses, and muscle contraction.	Regulated by aldosterone, insulin, and renal excretion.
Calcium (Ca^{2+})	Important for bone health, muscle contraction, nerve function, and blood clotting.	Controlled by Parathyroid Hormone (PTH), calcitonin, and vitamin D.
Magnesium ($Mg^{2+)}$	Vital for enzyme activity, muscle and nerve function, and energy production.	Regulated by renal excretion and reabsorption, PTH.
Chloride (Cl^-)	A major anion in ECF, helps maintain osmotic pressure and acid-base balance.	Regulated in conjunction with sodium by aldosterone.
Bicarbonate ($HCO_{3-)}$	A key component of the buffer system, helps maintain acid-base balance.	Regulated by the kidneys *via* reabsorption and secretion.
Phosphate ($PO^{4+}{}_{3-}$)	Important for energy production, bone health, and acid-base balance.	Regulated by PTH, vitamin D, and renal excretion.
Acid-Base Balance	Maintenance of pH within a narrow range (7.35-7.45), critical for normal cellular function.	Buffers (bicarbonate, phosphate, proteins), respiratory control of CO_2, renal regulation of H+ and HCO_3^-.
Renal Regulation	Kidneys play a crucial role in maintaining fluid and electrolyte balance by filtering blood, reabsorbing needed substances, and excreting waste and excess electrolytes.	Aldosterone increases Na+ reabsorption and K^+ excretion; ADH increases water reabsorption; natriuretic peptides promote Na^+ and water excretion.
Hormonal Control	Hormones like ADH, aldosterone, and PTH regulate fluid and electrolyte balance.	ADH increases water reabsorption; aldosterone increases Na^+ reabsorption and K^+ excretion; PTH increases Ca^{2+} reabsorption and $PO4^{3-}$ excretion.
Thirst Mechanism	Regulated by the hypothalamus, stimulates fluid intake in response to increased osmolality or decreased blood volume.	Activated by osmoreceptors in the hypothalamus and baroreceptors in the cardiovascular system.

(Table 1) cont.....

Key Concept	Description	Mechanisms and Regulation
Electrolyte Imbalances	Can lead to clinical conditions like hypernatremia, hyponatremia, hyperkalemia, hypokalemia, hypercalcemia, and hypocalcemia.	Diagnosed through blood tests; treated by addressing underlying causes, dietary modifications, medications, or intravenous fluids and electrolytes.
Dehydration	Loss of water exceeds intake, leading to reduced total body water.	Causes include excessive sweating, vomiting, diarrhea, and inadequate fluid intake. Treatment involves rehydration through oral or intravenous fluids.
Overhydration	Excessive fluid intake or retention leading to increased total body water.	Causes include renal failure, heart failure, and excessive IV fluid administration. Treatment involves fluid restriction and addressing the underlying cause.
Edema	Excess fluid accumulation in the interstitial spaces.	Causes include increased capillary pressure, decreased plasma protein levels, increased capillary permeability, and lymphatic obstruction. Treatment involves addressing the underlying cause.

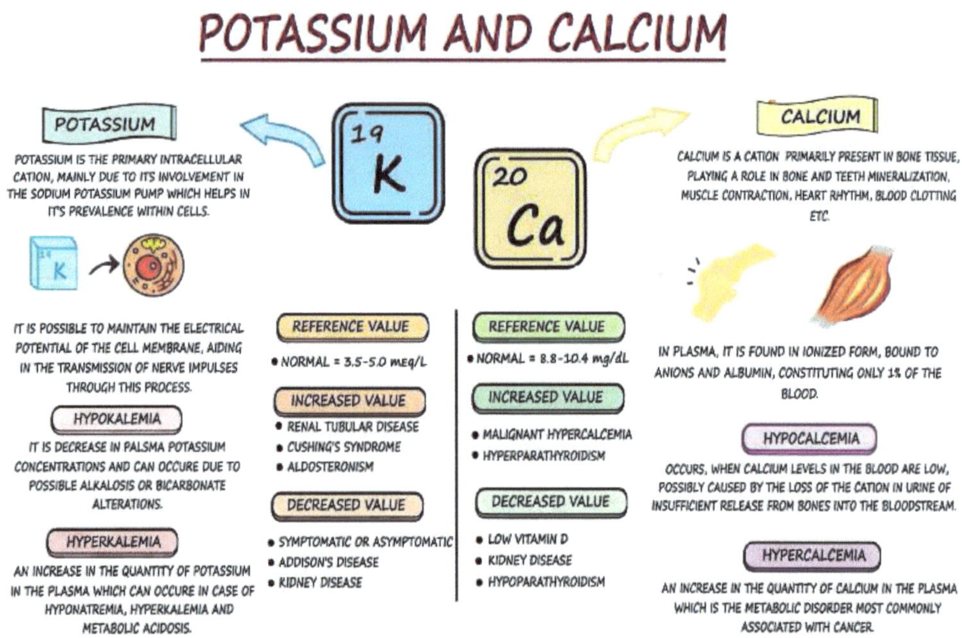

Fig. (1). Potassium and calcium.

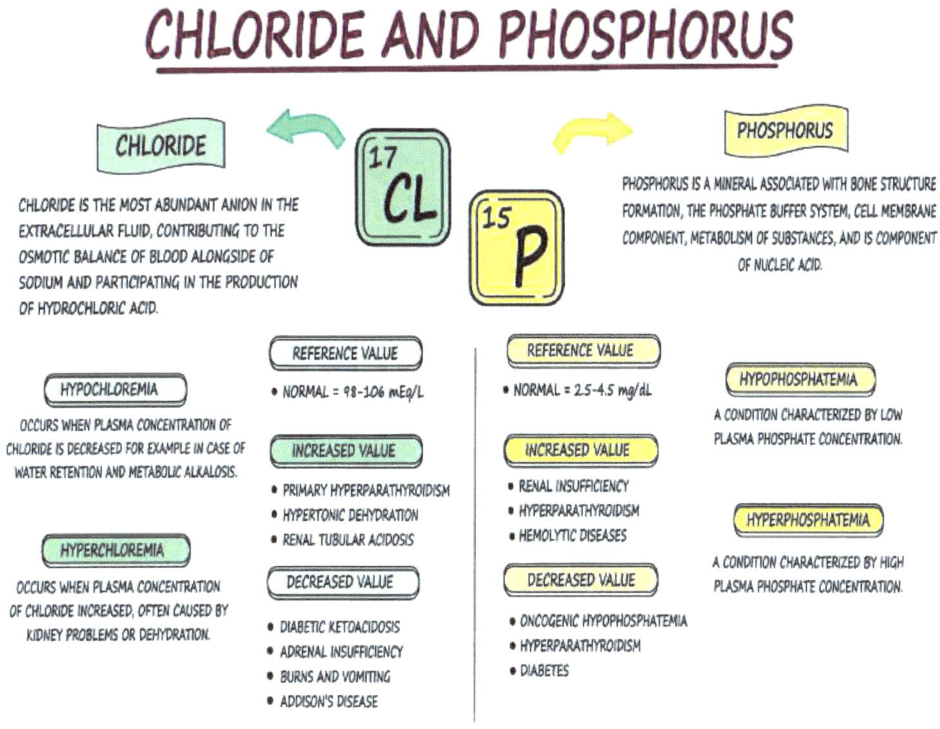

Fig. (2). Chloride and Phosphorus.

APPLIED ASPECTS OF HOMEOSTASIS OF FLUID AND ELECTROLYTE BALANCE

Causes and Consequences of Fluid and Electrolyte Disorders (Table 2)

Table 2. Causes and consequences of fluid and electrolyte disorders.

Aspect	Clinical Importance	Assessment Methods	Implications
Hydration Status	Crucial for diagnosing dehydration or overhydration.	Monitor weight, skin turgor, mucous membranes, urine output, and thirst.	Imbalances can lead to hypovolemia, hypervolemia, and affect cardiovascular and renal functions.
Electrolyte Levels	Important for diagnosing and treating electrolyte imbalances, such as hyperkalemia, hypokalemia, hypernatremia, *etc.*	Serum electrolyte tests (Na^+, K^+, Ca^{2+}, Mg^{2+}, Cl^-), and blood gas analysis.	Imbalances can cause arrhythmias, muscle weakness, neurological issues, and acid-base disturbances. (Fig. **1**, **2**).

Aspect	Clinical Importance	Assessment Methods	Implications
Blood Pressure	Indicates fluid status and helps diagnose hypertension or hypotension.	Blood pressure measurements using a sphygmomanometer.	Abnormalities may indicate fluid overload, dehydration, or cardiovascular issues.
Renal Function	Key in maintaining fluid and electrolyte balance and assessing kidney health.	Measure serum creatinine, Blood Urea Nitrogen (BUN), and Glomerular Filtration Rate (GFR).	Renal dysfunction can lead to imbalances and conditions like acute kidney injury and chronic kidney disease.
Hormonal Control	ADH, aldosterone, and natriuretic peptides play critical roles in fluid and electrolyte homeostasis.	Hormone level assays (ADH, aldosterone, renin, ANP).	Hormonal imbalances can lead to conditions like diabetes insipidus, SIADH, and hyperaldosteronism.
Acid-Base Balance	Essential for diagnosing and managing metabolic and respiratory acidosis/alkalosis.	Blood gas analysis (pH, pCO_2, HCO_3^-), serum anion gap.	Imbalances can lead to severe conditions affecting the respiratory and metabolic systems.
Cardiovascular Function	Fluid balance affects cardiac output and blood pressure.	Monitor heart rate, Jugular Venous Pressure (JVP), and Central Venous Pressure (CVP).	Fluid overload or deficit can impact cardiac function, leading to heart failure or shock.
Edema	Indicates fluid overload and potential heart, liver, or kidney issues.	Physical examination for pitting edema, lung auscultation for crackles.	Identifying the underlying cause is crucial for effective treatment.
Neurological Function	Electrolyte imbalances can affect brain function, leading to confusion, seizures, or coma.	Assess mental status, neurological exams, and electrolyte levels.	Prompt correction of imbalances is vital to prevent long-term neurological damage.

KEY CONCEPTS: EXTRACELLULAR AND INTRACELLULAR FLUID VOLUMES

Laboratory Tests Used To Diagnose Fluid And Electrolyte Imbalances (Table 3-5) [1-4]

Table 3. Extracellular and intracellular fluid.

Fluid Compartment	Description	Volume	Functions
Extracellular Fluid (ECF)	Comprises plasma (intravascular) and interstitial fluid (between cells).	About 20% of body weight	Maintains blood pressure, tissue perfusion, and nutrient/waste exchange between cells and blood vessels.

(Table 3) cont.....

Fluid Compartment	Description	Volume	Functions
Intracellular Fluid (ICF)	Fluid within cells.	About 40% of body weight	Vital for cellular function, metabolic processes, and maintaining cell shape and integrity.

APPLIED ASPECTS OF EXTRACELLULAR AND INTRACELLULAR FLUID VOLUMES

Table 4. Laboratory tests used to diagnose fluid and electrolyte imbalances.

Fluid Compartment	Clinical Importance	Clinical Assessments	Implications
Extracellular Fluid (ECF)	Key in diagnosing and managing fluid imbalances such as dehydration, overhydration, and edema.	Assess hydration status, blood pressure, jugular venous pressure, urine output, and rapid weight changes.	ECF volume abnormalities can lead to conditions like hypovolemia, hypervolemia, and shock, affecting cardiovascular health.
Intracellular Fluid (ICF)	Essential in understanding cellular health, osmotic balance, and the impact of electrolyte imbalances.	Measure serum electrolytes (Na^+, K^+), assess acid-base balance, and monitor clinical symptoms of cellular dysfunction.	ICF volume changes can affect cellular metabolism, leading to conditions like hyponatremia, hypernatremia, and cellular edema.

APPLIED ASPECTS OF CLINICAL ASSESSMENT OF EXTRACELLULAR FLUID (ECF) VOLUME

Table 5. Laboratory tests to diagnose fluid and electrolyte imbalances.

Aspect	Clinical Importance	Assessment Methods	Implications
Hydration Status	Vital for diagnosing dehydration or fluid overload.	Monitor skin turgor, mucous membranes, thirst, and daily weight changes.	Dehydration can lead to hypovolemia and shock, while fluid overload can cause edema and hypertension.
Blood Pressure	Indicates ECF volume status.	Measure using a sphygmomanometer.	Hypotension may indicate volume depletion, while hypertension may suggest volume overload.
Heart Rate	Reflects body's compensatory response to ECF volume changes.	Measure pulse rate.	Tachycardia may indicate hypovolemia, while bradycardia might suggest severe fluid overload or heart failure.

(Table 5) cont.....

Aspect	Clinical Importance	Assessment Methods	Implications
Jugular Venous Pressure (JVP)	Assesses intravascular volume status and central venous pressure.	Physical examination of neck veins.	An elevated JVP can indicate fluid overload, while a low JVP can suggest volume depletion.
Edema	Indicates fluid accumulation in the interstitial space.	Physical examination for pitting edema, especially in the lower extremities.	The Presence of edema can suggest heart failure, renal dysfunction, or liver disease.
Weight Changes	Rapid weight gain or loss can indicate ECF volume changes.	Daily weight measurements.	Rapid weight gain suggests fluid retention, while weight loss can indicate fluid loss.
Skin Turgor	Assesses hydration status by evaluating skin elasticity.	Pinch test on the skin (*e.g.*, forearm, abdomen).	Decreased skin turgor is a sign of dehydration.
Mucous Membranes	Moisture level reflects hydration status.	Examination of the oral mucosa.	Dry mucous membranes indicate dehydration.
Urine Output	Reflects kidney function and ECF volume status.	Measure urine output and perform urinalysis.	Oliguria can indicate volume depletion or renal dysfunction, while polyuria can suggest diabetes or diuretic use.
Laboratory Tests	Provide quantitative data on ECF volume and electrolyte balance.	Serum electrolytes (Na^+, K^+, Cl^-), BUN, creatinine, hematocrit, serum osmolality.	Imbalances in electrolytes can indicate ECF volume changes and potential renal or endocrine disorders.
Central Venous Pressure (CVP)	Direct measurement of pressure in the thoracic vena cava, reflecting ECF volume status.	Measurement *via* a central venous catheter.	An elevated CVP suggests fluid overload, while a low CVP indicates volume depletion.
Chest X-ray	Assesses for pulmonary congestion or edema, indicative of ECF volume overload.	Radiographic imaging.	Pulmonary congestion or pleural effusion can indicate fluid overload and heart failure.
Capillary Refill Time	Indicates peripheral perfusion and volume status.	Capillary refill test (pressing on a fingernail and observing the time it takes to return to normal color).	Prolonged capillary refill time suggests volume depletion and poor perfusion.
Electrolyte Imbalance	Important for detecting conditions like hyponatremia, hypernatremia, hypokalemia, and hyperkalemia.	Blood tests for serum electrolytes.	Electrolyte imbalances can lead to neuromuscular and cardiovascular symptoms, requiring prompt correction.
Neurological Symptoms	Electrolyte imbalances and severe ECF volume changes can affect neurological function.	Assess mental status and perform neurological exams.	Confusion, seizures, and coma can result from severe imbalances and volume changes.

Key Concepts Of Hormonal Control Of Fluid And Electrolyte Balance

Clinical Knowledge To Interpret Laboratory Results In The Context of Patient Symptoms (Table 6, 7) [1-4]

Table 6. Renin-Angiotensin-Aldosterone System (RAAS).

Hormone	Source	Function
Renin	Kidneys (juxtaglomerular cells)	Released in response to low blood pressure or low sodium levels. Converts angiotensinogen to angiotensin I in the blood.
Angiotensin II	Formed from Angiotensin I (*via* ACE in lungs)	Potent vasoconstrictor, increases blood pressure, stimulates aldosterone release, promotes sodium and water reabsorption in the kidneys.
Aldosterone	Adrenal cortex	Promotes sodium reabsorption and potassium excretion in the kidneys, increasing water reabsorption and blood volume.

Table 7. Antidiuretic Hormone (ADH).

Hormone	Source	Function
ADH (Vasopressin)	Hypothalamus (stored and released by the posterior pituitary)	Increases water reabsorption in the kidneys by inserting aquaporin channels in the collecting ducts, reducing urine output and increasing blood volume.

APPLIED ASPECTS OF HORMONAL CONTROL OF FLUID AND ELECTROLYTE BALANCE

Impact of Fluid And Electrolyte Disorders On Overall Health And Disease Outcomes (Table 8-11)

Table 8. Applied aspects of hormonal control: Renin-Angiotensin-Aldosterone System (RAAS).

Hormone	Clinical Importance	Assessment Methods	Implications
Renin	Key in diagnosing and managing conditions like hypertension and renal artery stenosis.	Blood renin activity (PRA) or concentration (PRC) tests.	Elevated renin levels can indicate renal artery stenosis or low blood volume; low levels can suggest primary hyperaldosteronism.
Angiotensin II	Important for managing blood pressure and fluid balance.	Assess indirectly *via* blood pressure response to ACE inhibitors or ARBs.	Overactivity can cause hypertension; inhibitors are used in treating hypertension and heart failure.

(Table 8) cont.....

Hormone	Clinical Importance	Assessment Methods	Implications
Aldosterone	Crucial for diagnosing and treating conditions like hyperaldosteronism and Addison's disease.	Serum aldosterone and Plasma Renin Activity (PRA) ratio, urinary aldosterone excretion.	High levels indicate hyperaldosteronism (Conn's syndrome); low levels can be seen in Addison's disease.

Table 9. Applied aspects of hormonal control: Antidiuretic Hormone (ADH).

Hormone	Clinical Importance	Assessment Methods	Implications
ADH (Vasopressin)	Essential in diagnosing and managing disorders like diabetes insipidus and SIADH (Syndrome of Inappropriate ADH).	Plasma ADH levels, water deprivation test, and response to desmopressin.	Low ADH levels or response indicate diabetes insipidus; high levels or inappropriate secretion indicate SIADH.

Table 10. Overview of functions and implications: hormones.

Hormone	Functions	Implications
Renin	Triggers the RAAS cascade, increasing blood pressure and volume.	Abnormal levels can lead to conditions like hypertension, hypotension, and electrolyte imbalances.
Angiotensin II	Causes vasoconstriction, stimulates aldosterone release, promotes sodium and water reabsorption in the kidneys.	Overactivity can lead to hypertension and heart failure; treated with ACE inhibitors or ARBs.
Aldosterone	Enhances sodium and water reabsorption, increases blood volume and pressure, promotes potassium excretion.	Imbalances can cause hyperaldosteronism or Addison's disease, impacting fluid and electrolyte balance.
ADH (Vasopressin)	Increases water reabsorption in the kidneys, reducing urine output and increasing blood volume.	Disorders include diabetes insipidus (low ADH) and SIADH (high ADH), affecting hydration and osmolality.

Table 11. Key concepts of the principles of correcting fluid and electrolyte losses.

Aspect	Details
Assessment of Fluid and Electrolyte Status	- Clinical Evaluation: History, physical examination (skin turgor, mucous membranes, vital signs), urine output. - Laboratory Tests: Serum electrolytes (Na^+, K^+, Cl^- HCO^{3-}), BUN, creatinine, glucose, ABGs, urine specific gravity.
Identify the Type and Cause of Imbalance	- Fluid Imbalance: Dehydration, overhydration. - Electrolyte Imbalance: Hyponatremia, hypernatremia, hypokalemia, hyperkalemia, others (calcium, magnesium).
Principles of Correction	- Fluid Replacement: Oral Rehydration (ORS), intravenous fluids (crystalloids, colloids). - Electrolyte Replacement: Sodium (hyponatremia, hypernatremia), potassium (hypokalemia, hyperkalemia), calcium, magnesium.

(Table 11) cont.....

Aspect	Details
Monitoring and Adjusting Therapy	- Continuous assessment of clinical status, vital signs, urine output, and laboratory tests. - Adjust fluid and electrolyte replacement based on ongoing losses and response to treatment.
Special Considerations	- Rate of Correction: Avoid rapid changes to prevent complications. - Underlying Conditions: Address root causes (*e.g.*, diabetes, infection) - Comorbidities: Consider the impact of concurrent medical conditions.
Prevention of Recurrence	- Patient education on adequate intake during illness or stress. - Long-term management for chronic conditions with dietary changes, medications, and follow-up.

BIOCHEMICAL BASIS, RELEVANT LAB TESTS, INTERPRETATION, AND APPLIED ASPECTS: DISORDERS OF FLUID AND ELECTROLYTE BALANCE

Biochemical Basis, Relevant Lab Tests, Analytical Methods, Interpretation and Applied Aspects (Table 12, 13) [1-4]

Table 12. Biochemical basis, relevant lab tests, interpretation, and applied aspects of fluid and electrolyte balance.

Condition	Biochemical Basis	Causes and Clinical Signs	Biochemical Changes	Relevant Laboratory Tests	Interpretation of Analytical Results	Analytical Methods
Dehydration	Loss of water leading to electrolyte imbalances	Vomiting, diarrhea, excessive sweating; signs of dehydration	Increased serum osmolality, electrolyte imbalances	Serum electrolytes (Na+, K+), BUN, creatinine, and urine specific gravity	Elevated serum sodium and potassium; increased BUN/creatinine ratio	Electrolyte analysis (flame photometry, ion-selective electrodes), and BUN/creatinine assays
Diabetes Insipidus	Insufficient ADH production or response	Central or nephrogenic causes; polyuria and polydipsia	Decreased urine osmolality despite high serum osmolality	Water deprivation test, urine and serum osmolality, and ADH levels	Inability to concentrate urine during the water deprivation test; elevated serum osmolality	Osmolality measurement (freezing point depression), and ADH assays
Extracellular Fluid (ECF) Volume Loss	Reduction in extracellular fluid volume	Fluid loss (*e.g.*, diarrhea, vomiting); signs of volume depletion	Increased serum osmolality, electrolyte imbalances	Serum electrolytes, BUN, creatinine, and urine specific gravity	Elevated serum sodium and potassium; increased BUN/creatinine ratio	Electrolyte analysis (flame photometry, ion-selective electrodes), and BUN/creatinine assays
Hypernatremia	Elevated serum sodium levels (>145 mEq/L)	Dehydration, excessive sodium intake; signs of hypernatremia	Increased serum sodium	Serum electrolytes and urine osmolality	Elevated serum sodium; assessed for underlying causes	Electrolyte analysis and osmolality measurement

(Table 12) cont.....

Condition	Biochemical Basis	Causes and Clinical Signs	Biochemical Changes	Relevant Laboratory Tests	Interpretation of Analytical Results	Analytical Methods
Hyperkalemia and Pseudohyperkalemia	Elevated serum potassium levels (>5.0 mEq/L)	Renal failure, potassium-sparing diuretics; signs of hyperkalemia	Increased serum potassium	Serum potassium and ECG changes	Elevated serum potassium; assessed for pseudohyperkalemia (hemolysis)	Potassium measurement and ECG evaluation
Hypokalemia	Decreased serum potassium levels (<3.5 mEq/L)	Diuretics, GI losses, alkalosis; signs of hypokalemia	Decreased serum potassium	Serum potassium and ECG changes	Decreased serum potassium; assessed for underlying causes	Potassium measurement and ECG evaluation
Hyponatremia and Pseudohyponatraemia	Decreased serum sodium levels (<135 mEq/L)	Fluid overload, SIADH, medications; signs of hyponatremia	Decreased serum sodium, hypoosmolality	Serum sodium and osmolality	Decreased serum sodium; assessed for true hyponatremia *vs.* pseudohyponatremia	Sodium measurement and osmolality calculation
Shock	Circulatory failure leading to inadequate tissue perfusion	Hypovolemic, cardiogenic, septic shock; signs of shock	Lactic acidosis, electrolyte imbalances	Blood lactate, electrolytes, and ABGs	Elevated lactate; metabolic acidosis	Lactate measurement and electrolyte analysis
Syndrome of Inappropriate Antidiuretic Hormone (SIADH)	Excessive release of ADH leading to water retention	CNS disorders, lung diseases, medications; signs of SIADH	Decreased serum sodium, concentrated urine	Serum sodium, osmolality, and urine osmolality	Decreased serum sodium; concentrated urine osmolality	Sodium and osmolality measurement and ADH assessment
Oncotic Disorders (*e.g.*, Edema, Ascites)	Abnormalities in plasma proteins affecting oncotic pressure	Liver disease, nephrotic syndrome; signs of edema, ascites	Hypoalbuminemia, decreased oncotic pressure	Serum albumin, total protein, and liver function tests	Decreased serum albumin; elevated liver enzymes	Albumin and total protein measurement and liver panel

RELEVANT LABORATORY TESTS, ANALYTICS METHODS AVAILABLE AND APPLIED ASPECTS

Summary

Maintaining fluid and electrolyte balance is crucial for normal physiological function and homeostasis. Disorders such as hyponatremia, hyperkalemia, and dehydration have serious clinical implications. Diagnosis involves clinical evaluation and laboratory tests, including serum electrolyte levels, osmolality, and renal function tests. Treatment strategies vary depending on the underlying cause and may include fluid replacement, dietary modifications, and pharmacological interventions. Understanding the biochemical and physiological basis of these disorders is essential for effective diagnosis, management, and prevention of complications.

Table 13. Relevant laboratory tests, analytics methods available, and applied aspects

Test	Analytical Methods	Applied Aspects
Aldosterone	- Immunoassay (ELISA, RIA) - Mass spectrometry (LC-MS/MS)	- Assesses primary aldosteronism (Conn's syndrome) - Helps diagnose secondary hyperaldosteronism (*e.g.*, heart failure, cirrhosis) - Used to evaluate electrolyte imbalance and hypertension due to aldosterone dysregulation.
Electrolytes (Sodium, Potassium, Chloride)	- Ion-Selective Electrodes (ISE) - Flame photometry	- Routine assessment of electrolyte imbalances - Identifies conditions like dehydration, acidosis/alkalosis, Addison's disease, or hyperkalemia/hypokalemia. - Monitors therapy in patients with heart failure, kidney disease, or metabolic disorders.
Serum and Urine Osmolality	- Freezing point depression osmometry - Vapor pressure osmometry	- Diagnosis of fluid balance disorders such as diabetes insipidus or SIADH - Assesses kidney function and urine concentrating ability - Helps differentiate between prerenal and renal causes of acute kidney injury.
Renin-Angiotensin Axis (RTA) Assessment	- Plasma Renin Activity (PRA) assay - Aldosterone-Renin Ratio (ARR) - LC-MS/MS	- Diagnoses renovascular hypertension and primary aldosteronism - Assesses the effectiveness of renin-angiotensin system inhibitors in hypertension treatment - Evaluates patients with unexplained hypokalemia and resistant hypertension.
Urine and Fecal Electrolytes	- Flame photometry - Ion-selective electrodes - Inductively Coupled Plasma Mass Spectrometry (ICP-MS)	- Assesses causes of chronic diarrhea and fluid loss - Differentiates between renal and extrarenal causes of electrolyte imbalances - Used in the investigation of conditions like cystic fibrosis, chronic kidney disease, and electrolyte-wasting syndromes.

EXERCISES

1. Examining the physiological mechanisms that govern sodium and water balance within the body.
2. Outlining the clinical features and laboratory findings associated with hyperkalemia.
3. Investigating the diagnosis and management of hyponatremia in hospitalized patients.
4. Exploring the kidneys' role in maintaining electrolyte balance.
5. Identifying common causes and treatment options for dehydration.
6. Understanding how diuretics can impact fluid and electrolyte balance.

7. Delving into the pathophysiology of hypernatremia and its clinical management.
8. Detailing the diagnostic approach for patients suspected of experiencing fluid overload.
9. Examining the influence of acid-base disorders on electrolyte balance.
10. Exploring the clinical significance of serum osmolality in assessing fluid and electrolyte disorders.

CONCLUSION

It is crucial to have a deep understanding of the biochemical and physiological aspects of fluid and electrolyte disorders in order to effectively manage them in clinical practice. The advancements in diagnostic methods and targeted treatments have greatly enhanced patient outcomes. Staying updated with the latest developments in the field through continuous research and education is imperative. Optimal patient care necessitates a comprehensive approach that combines clinical assessment with laboratory data.

REFERENCES

[1] Rout, P.; Badireddy, M. Hyponatremia. [Updated 2025 Dec 13]. In: StatPearls [Internet]. Treasure Island (FL): StatPearls Publishing. **2025**. Available from: https://www.ncbi.nlm.nih.gov/books/NBK470386/

[2] Mount, D.B. *Fluid and electrolyte disturbances. Schwartz's Principles of Surgery,* 11th ed; McGraw Hill Education, **2019**.

[3] Fujimaru, T.; Hirose, K.; Yazawa, M.; Nagahama, M.; Kovesdy, C.P. Electrolyte Winter Seminar Collaborative Group. Pathophysiology and causes of hyperkalemia: unraveling causes beyond kidney dysfunction *Clin Exp Nephrol. 2025,* **2025**, *29*(11), 1524-1534. [http://dx.doi.org/10.1007/s10157-025-02711-x]

[4] Hopkins, E.; Sanvictores, T.; Sharma, S. Physiology, Acid Base Balance. [Updated 2022 Sep 12]. In: StatPearls [Internet]. Treasure Island (FL): StatPearls Publishing. **2025**. Available from: https://www.ncbi.nlm.nih.gov/books/NBK507807/

CHAPTER 8

Gastrointestinal and Pancreatic Disease

Abstract: The field of gastrointestinal and pancreatic diseases encompasses a wide range of conditions that have significant clinical implications. This chapter delves into the biochemical and physiological aspects of the Gastrointestinal (GI) tract and pancreas, offering insights into the pathophysiology, diagnosis, and management of common disorders such as inflammatory bowel disease, pancreatitis, and pancreatic cancer. Recent advancements in laboratory medicine, including the utilization of specific biomarkers, have led to improved diagnostic accuracy and therapeutic strategies. The emphasis is on integrating laboratory results with clinical findings to improve patient outcomes.

LEARNING OBJECTIVES

After reading this chapter, one should:

1. Be able to provide a general overview and describe the functions of the gastrointestinal and pancreatic systems.
2. Be able to describe diseases of gastrointestinal and pancreatic function.
3. Be able to recognize the causes, clinical signs, and symptoms of the following disorders relating to gastrointestinal and pancreatic function.
4. Be able to describe which laboratory investigations are important in the detection, diagnosis, and management of gastrointestinal and pancreatic diseases.
5. Be able to describe the following specific laboratory investigations important to the study of disorders of gastrointestinal and pancreatic function.
6. Be able to describe analytical methods available for their measurement.
7. Be able to select laboratory investigations and interpret the analytical results in the context of clinical signs and symptoms.

KEY CONCEPTS: GASTROINTESTINAL AND PANCREATIC SYSTEMS

Functions of the Gastrointestinal and Pancreatic Systems (Table 1-4; Fig. 1, Fig. 2). [1, 2]

Simmi Kharb

Table 1. Functions of the gastrointestinal and pancreatic systems.

System Component	General Overview	Functions
Mouth	Entry point of the GI system	Mechanical breakdown of food (chewing) and enzymatic digestion (saliva).
Esophagus	A muscular tube connecting the mouth to the stomach	Transports food *via* peristalsis.
Stomach	A muscular organ that mixes food with gastric juices	Protein digestion, food conversion to chyme, secretion of hydrochloric acid, and digestive enzymes.
Small Intestine	A long, coiled tube divided into the duodenum, jejunum, and ileum	A major site for digestion and nutrient absorption, the secretion of digestive enzymes, and bile interaction.
Large Intestine	Shorter, wider tube connected to the small intestine	Absorbs water and electrolytes, forms and expels feces.
Rectum and Anus	Terminal parts of the GI system	Store and control expulsion of feces.
Pancreas	Gland with both endocrine and exocrine functions	Exocrine: Produces digestive enzymes and bicarbonate for digestion. Endocrine: Produces hormones (insulin, glucagon) to regulate blood glucose.

KEY CONCEPTS: INTESTINAL ABSORPTION OF PROTEINS, FATS, AND CARBOHYDRATES

Table 2. Functions of the gastrointestinal and pancreatic systems.

Nutrient	Digestive Process	Absorption Mechanism	Transport Pathway	Key Enzymes and Components
Proteins	Proteins are broken down into peptides and amino acids	Amino acids and small peptides are absorbed primarily through active transport and facilitated diffusion	Amino acids and peptides enter the enterocytes (intestinal cells) and are then transported into the bloodstream *via* the portal vein to the liver	Pepsin (stomach), trypsin, chymotrypsin, carboxypeptidase (pancreas), and peptidases (small intestine)

(Table 2) cont.....

Nutrient	Digestive Process	Absorption Mechanism	Transport Pathway	Key Enzymes and Components
Fats	Fats are emulsified by bile salts and digested into fatty acids and monoglycerides	Fatty acids and monoglycerides form micelles with bile salts, which diffuse into enterocytes. Inside enterocytes, they are re-esterified to form triglycerides and packaged into chylomicrons	Chylomicrons enter the lymphatic system *via* lacteals and eventually enter the bloodstream. Short- and medium-chain fatty acids are absorbed directly into the bloodstream	Lipase (pancreas), bile salts (liver), and colipase (pancreas)
Carbohydrates	Carbohydrates are broken down into monosaccharides (glucose, fructose, galactose)	Monosaccharides are absorbed through active transport (glucose and galactose *via* SGLT1) and facilitated diffusion (fructose *via* GLUT5)	Monosaccharides enter enterocytes and are then transported into the bloodstream *via* the portal vein to the liver	Amylase (saliva and pancreas), maltase, sucrase, and lactase (small intestine)

KEY CONCEPTS: ENDOCRINE AND EXOCRINE FUNCTIONS OF THE PANCREAS [3, 4]

Table 3. Functions of the pancreatic systems.

Function	Component	Hormones/Enzymes	Primary Actions	Regulation and Control
Endocrine	Islets of Langerhans	**Insulin**	Lowers blood glucose by promoting cellular glucose uptake, glycogenesis, and lipogenesis	Stimulated by high blood glucose levels (hyperglycemia)
		Glucagon	Raises blood glucose by stimulating glycogenolysis and gluconeogenesis	Stimulated by low blood glucose levels (hypoglycemia)
		Somatostatin	Inhibits the release of both insulin and glucagon and slows down digestive processes	Regulated by blood glucose levels and other hormones
		Pancreatic Polypeptide	Regulates pancreatic secretions and influences hepatic glycogen levels	Regulated by food intake, blood glucose levels, and the autonomic nervous system

(Table 3) cont.....

Function	Component	Hormones/Enzymes	Primary Actions	Regulation and Control
Exocrine	Acinar Cells	**Amylase**	Breaks down carbohydrates into simple sugars	Secretion stimulated by Cholecystokinin (CCK) and the vagus nerve (Fig. 1)
		Lipase	Breaks down fats into fatty acids and glycerol	Secretion stimulated by CCK and the vagus nerve (Fig. 2)
		Proteases (Trypsin, Chymotrypsin, Carboxypeptidase)	Break down proteins into peptides and amino acids	Secretion stimulated by CCK and the vagus nerve
		Nucleases (DNase, RNase)	Break down nucleic acids into nucleotides	Secretion stimulated by CCK and the vagus nerve
	Duct Cells	**Bicarbonate**	Neutralizes stomach acid in the small intestine	Secretion stimulated by secretin

Table 4. Functions of the gastrointestinal system: Vitamin B_{12}.

Step	Location	Process	Key Components
Release from Food	Stomach	Vitamin B12 is released from food proteins by gastric acid and pepsin.	Gastric acid and pepsin
Binding to R-Protein	Stomach	Vitamin B12 binds to R-protein (haptocorrin) secreted by salivary glands and gastric mucosa.	R-protein (haptocorrin)
Transport to Small Intestine	Stomach to Duodenum	The Vitamin B12-R-protein complex moves to the duodenum.	Peristalsis
Release from R-Protein	Duodenum	Pancreatic proteases degrade R-protein, releasing Vitamin B12.	Pancreatic proteases
Binding to Intrinsic Factor (IF)	Duodenum	Free Vitamin B12 binds to intrinsic factor, a glycoprotein produced by gastric parietal cells.	Intrinsic Factor (IF)
Absorption	Ileum	The Vitamin B12-IF complex binds to specific receptors on enterocytes in the ileum and is absorbed.	Intrinsic factor receptors and enterocytes
Transport in Blood	Enterocytes to the Bloodstream	Inside enterocytes, Vitamin B12 is released from IF and binds to transcobalamin II for transport in the bloodstream.	Transcobalamin II
Storage and Utilization	Liver, Tissues	Vitamin B12 is stored in the liver and used in various tissues for DNA synthesis and neurological function.	Liver and body tissues

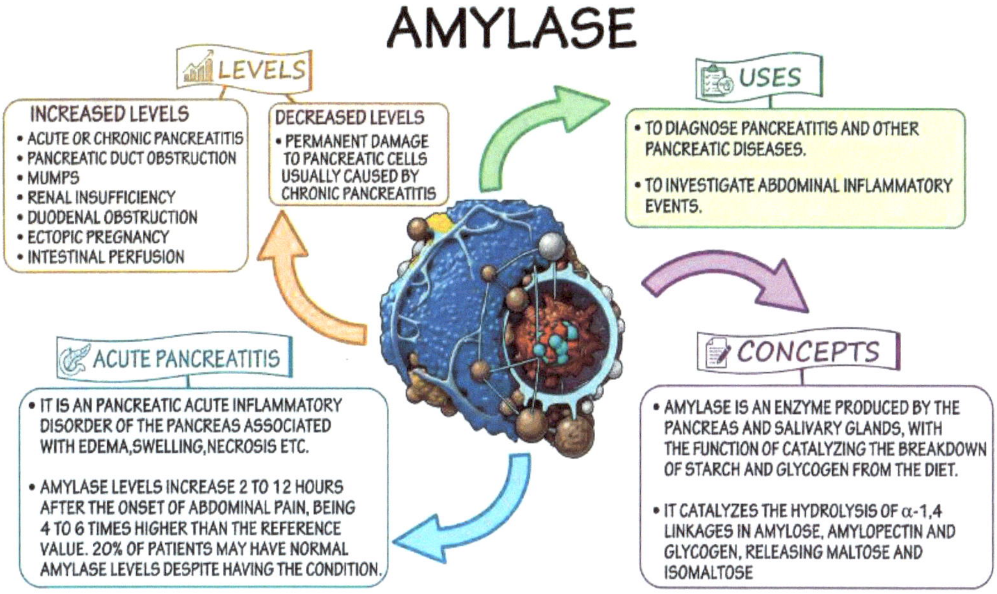

Fig. (1). Amylase.

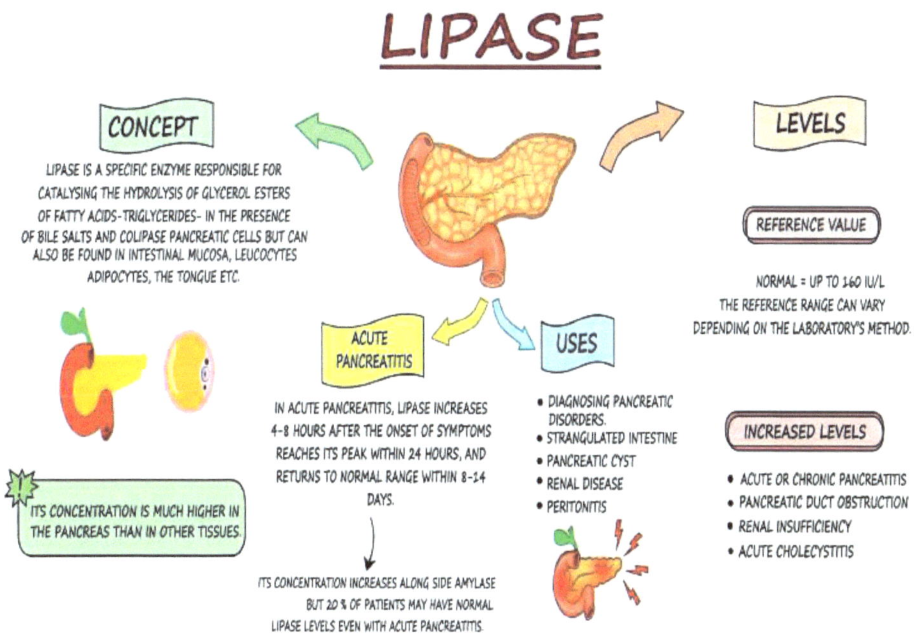

Fig. (2). Lipase.

KEY CONCEPTS: VITAMIN B_{12} ABSORPTION

Biochemical Basis, Relevant LAB Tests, Interpretation, and Applied Aspects

Laboratory Investigations in Detection, Diagnosis, and Management of Gastrointestinal and Pancreatic Diseases (Table 5, 6) [1 - 4]

Table 5. Laboratory investigations in the detection, diagnosis, and management of gastrointestinal and pancreatic diseases.

Disease/Disorder	Biochemical Basis	Causes	Clinical Signs and Symptoms	Biochemical Changes
Acute Pancreatitis	Activation of pancreatic enzymes within the pancreas, leading to autodigestion and inflammation	Gallstones, alcohol abuse, hypertriglyceridemia, and certain medications	Severe abdominal pain, nausea, vomiting, fever, tachycardia, and abdominal tenderness	Elevated serum amylase and lipase, elevated liver enzymes, and hypocalcemia
Chronic Pancreatitis	Persistent inflammation causing irreversible damage to pancreatic tissue	Chronic alcohol use, genetic mutations, and autoimmune conditions	Chronic abdominal pain, weight loss, steatorrhea, and diabetes mellitus	Elevated serum amylase and lipase (may be normal in late stages), and pancreatic calcifications visible on imaging
Autoimmune Bowel Disease	Immune-mediated inflammation targeting the gastrointestinal tract	Genetic predisposition, environmental triggers, and immune dysregulation	Diarrhea, abdominal pain, weight loss, fatigue, anemia, and extraintestinal manifestations (*e.g.*, arthritis, uveitis)	Elevated inflammatory markers (CRP, ESR), presence of autoantibodies, and abnormal fecal calprotectin
Causes of Gastric Ulceration	Breakdown of the gastric mucosal barrier leading to erosion and ulcer formation in the stomach lining	Helicobacter pylori infection, chronic NSAID use, stress, and smoking	Epigastric pain, nausea, vomiting, weight loss, bloating, hematemesis, and melena	Positive urea breath test or stool antigen test for *H. pylori* and elevated gastrin levels in Zollinger-Ellison syndrome

(Table 5) cont.....

Disease/Disorder	Biochemical Basis	Causes	Clinical Signs and Symptoms	Biochemical Changes
Inflammatory Bowel Disease (IBD)	Chronic inflammation of the gastrointestinal tract, primarily Crohn's disease and ulcerative colitis	Genetic predisposition, environmental factors, and immune response	Diarrhea, abdominal pain, rectal bleeding, weight loss, fatigue, and extraintestinal manifestations	Elevated CRP and ESR, anemia, hypoalbuminemia, leukocytosis, abnormal fecal calprotectin, and endoscopic findings of inflammation
Neuroendocrine Tumors	Tumors arising from neuroendocrine cells, leading to excess hormone production and associated symptoms	Genetic mutations (*e.g.*, MEN1), sporadic cases, and familial syndromes	Symptoms vary depending on the hormone produced: flushing, diarrhea, wheezing (carcinoid syndrome), hypoglycemia, and ulcers	Elevated specific hormone levels (*e.g.*, serotonin, chromogranin A, insulin, gastrin), imaging studies, and biopsy findings

Table 6. Laboratory investigations in the detection, diagnosis, and management of gastrointestinal and pancreatic diseases.

Laboratory Test	Detection, Diagnosis, and Management/Monitoring	Interpretation of Analytical Results	Analytical Methods
$^{14}CO_2$ Urea Breath Test for *H. pylori*	Detection of *H. pylori* infection	Positive results indicate active *H. pylori* infection, common in patients with peptic ulcers.	Urea breath test with $^{14}CO_2$ isotope
Amylase and Macroamylase	Diagnosis of pancreatitis and macroamylasemia	Elevated serum amylase suggests acute pancreatitis; the presence of macroamylase suggests macroamylasemia.	Enzymatic colorimetric assay
CA 19-9	Monitoring of pancreatic cancer	Elevated levels indicate pancreatic cancer, though they may also rise in other conditions like cholangitis.	Immunoassay

(Table 6) cont.....

Laboratory Test	Detection, Diagnosis, and Management/Monitoring	Interpretation of Analytical Results	Analytical Methods
Carcinoembryonic Antigen (CEA)	Monitoring colorectal cancer and other cancers	Elevated levels suggest malignancy; used primarily for monitoring treatment response and recurrence.	Enzyme-Linked Immunosorbent Assay (ELISA)
Chromogranin A	Diagnosis and monitoring of neuroendocrine tumors	Elevated levels indicate the presence of neuroendocrine tumors; used for monitoring treatment response.	Immunoassay
Elastase	Assessment of pancreatic exocrine function	Low fecal elastase indicates pancreatic insufficiency.	Enzyme-Linked Immunosorbent Assay (ELISA)
Endomysial Autoantibodies	Diagnosis of celiac disease	Positive results indicate celiac disease.	Immunofluorescence
Fecal Calprotectin	Detection and monitoring of Inflammatory Bowel Disease (IBD)	Elevated levels indicate active inflammation in the gastrointestinal tract, commonly seen in IBD.	Enzyme-Linked Immunosorbent Assay (ELISA)
Fecal Occult Blood	Screening for colorectal cancer	Positive results indicate the presence of blood in stool, suggesting gastrointestinal bleeding.	Guaiac-based or immunochemical Fecal Occult Blood Test (iFOBT)
Colon Cancer Screening Methods	Detection of colorectal cancer	Positive results on screening tests indicate possible colorectal cancer.	Colonoscopy, sigmoidoscopy, Fecal Occult Blood Test (FOBT)
Hydrogen Breath Test for Lactose Intolerance	Diagnosis of lactose intolerance	Elevated hydrogen levels indicate lactose malabsorption and intolerance.	Hydrogen breath test
IgE and Specific IgE	Diagnosis of allergic reactions and monitoring allergy treatment	Elevated total and specific IgE levels indicate allergic sensitization.	Immunoassay

(Table 6) cont.....

Laboratory Test	Detection, Diagnosis, and Management/Monitoring	Interpretation of Analytical Results	Analytical Methods
Intrinsic Factor Antibodies	Diagnosis of pernicious anemia	Positive results indicate the presence of antibodies against intrinsic factor, leading to Vitamin B12 deficiency.	Enzyme-Linked Immunosorbent Assay (ELISA)
Lipase	Diagnosis of pancreatitis	Elevated serum lipase indicates acute pancreatitis; more specific than amylase.	Enzymatic colorimetric assay
Serotonin and 5-Hydroxyindoleacetic Acid (5-HIAA)	Diagnosis and monitoring of carcinoid syndrome	Elevated urine 5-HIAA and serum serotonin levels indicate carcinoid syndrome.	High-Performance Liquid Chromatography (HPLC)
Transglutaminase Autoantibodies	Diagnosis of celiac disease	Positive results indicate celiac disease.	Enzyme-Linked Immunosorbent Assay (ELISA)
Vitamin B12 and Folate	Diagnosis of deficiency states	Low levels of Vitamin B12 and folate indicate deficiency, leading to anemia and neurological symptoms.	Chemiluminescence immunoassay
Xylose Absorption Test	Assessment of carbohydrate absorption in the small intestine	Low urinary xylose excretion indicates malabsorption.	Colorimetric assay
Stimulation/Suppression Tests	Assessment of endocrine function (*e.g.*, adrenal, thyroid)	Changes in hormone levels in response to stimulation or suppression indicate endocrine gland function or dysfunction.	Various, depending on the specific hormone being tested (*e.g.*, ACTH stimulation test, dexamethasone suppression test)

RELEVANT LABORATORY TESTS FOR DETECTION, DIAGNOSIS, MANAGEMENT, AND MONITORING

Relevant Laboratory Tests, Analytics Methods Available and Applied Aspects

Analytical Methods Available for their Measurement (Table 7)

Table 7. Analytical methods available for their measurement.

Laboratory Test	Detection, Diagnosis, and Management/Monitoring	Interpretation of Analytical Results	Analytical Methods
$^{14}CO_2$ Urea Breath Test for *H. pylori*	Detection of *H. pylori* infection	Positive results indicate active *H. pylori* infection, common in patients with peptic ulcers.	Urea breath test with $^{14}CO_2$ isotope
Amylase and Macroamylase	Diagnosis of pancreatitis and macroamylasemia	Elevated serum amylase suggests acute pancreatitis; the presence of macroamylase suggests macroamylasemia.	Enzymatic colorimetric assay
CA 19-9	Monitoring of pancreatic cancer	Elevated levels indicate pancreatic cancer, though they may also rise in other conditions like cholangitis.	Immunoassay
Carcinoembryonic Antigen (CEA)	Monitoring colorectal cancer and other cancers	Elevated levels suggest malignancy; used primarily for monitoring treatment response and recurrence.	Enzyme-Linked Immunosorbent Assay (ELISA)
Chromogranin A	Diagnosis and monitoring of neuroendocrine tumors	Elevated levels indicate the presence of neuroendocrine tumors; used for monitoring treatment response.	Immunoassay
Elastase	Assessment of pancreatic exocrine function	Low fecal elastase indicates pancreatic insufficiency.	Enzyme-Linked Immunosorbent Assay (ELISA)
Endomysial Autoantibodies	Diagnosis of celiac disease	Positive results indicate celiac disease.	Immunofluorescence
Fecal Calprotectin	Detection and monitoring of Inflammatory Bowel Disease (IBD)	Elevated levels indicate active inflammation in the gastrointestinal tract, commonly seen in IBD.	Enzyme-Linked Immunosorbent Assay (ELISA)

(Table 7) cont.....

Laboratory Test	Detection, Diagnosis, and Management/Monitoring	Interpretation of Analytical Results	Analytical Methods
Fecal Occult Blood	Screening for colorectal cancer	Positive results indicate the presence of blood in stool, suggesting gastrointestinal bleeding.	Guaiac-based or immunochemical Fecal Occult Blood Test (iFOBT)
Colon Cancer Screening Methods	Detection of colorectal cancer	Positive results on screening tests indicate possible colorectal cancer.	Colonoscopy, sigmoidoscopy, Fecal Occult Blood Test (FOBT)
Hydrogen Breath Test for Lactose Intolerance	Diagnosis of lactose intolerance	Elevated hydrogen levels indicate lactose malabsorption and intolerance.	Hydrogen breath test
IgE and Specific IgE	Diagnosis of allergic reactions and monitoring allergy treatment	Elevated total and specific IgE levels indicate allergic sensitization.	Immunoassay
Intrinsic Factor Antibodies	Diagnosis of pernicious anemia	Positive results indicate the presence of antibodies against intrinsic factor, leading to Vitamin B12 deficiency.	Enzyme-Linked Immunosorbent Assay (ELISA)
Lipase	Diagnosis of pancreatitis	Elevated serum lipase indicates acute pancreatitis; more specific than amylase.	Enzymatic colorimetric assay
Serotonin and 5-Hydroxyindoleacetic Acid (5-HIAA)	Diagnosis and monitoring of carcinoid syndrome	Elevated urine 5-HIAA and serum serotonin levels indicate carcinoid syndrome.	High-Performance Liquid Chromatography (HPLC)
Transglutaminase Autoantibodies	Diagnosis of celiac disease	Positive results indicate celiac disease.	Enzyme-Linked Immunosorbent Assay (ELISA)
Vitamin B12 and Folate	Diagnosis of deficiency states	Low levels of Vitamin B12 and folate indicate deficiency, leading to anemia and neurological symptoms.	Chemiluminescence immunoassay
Xylose Absorption Test	Assessment of carbohydrate absorption in the small intestine	Low urinary xylose excretion indicates malabsorption.	Colorimetric assay

(Table 7) cont.....

Laboratory Test	Detection, Diagnosis, and Management/Monitoring	Interpretation of Analytical Results	Analytical Methods
Stimulation/Suppression Tests	Assessment of endocrine function (*e.g.*, adrenal, thyroid)	Changes in hormone levels in response to stimulation or suppression indicate endocrine gland function or dysfunction.	Various, depending on the specific hormone being tested (*e.g.*, ACTH stimulation test, dexamethasone suppression test)

SUMMARY

Gastrointestinal and pancreatic diseases are common and can lead to serious complications if not properly managed. Conditions like IBD, pancreatitis, and pancreatic cancer require a comprehensive understanding for diagnosis and treatment. Diagnosis involves clinical assessment, lab tests, imaging, and biomarker analysis. Management varies but may include medication, dietary changes, and surgery. Biomarker research has improved disease diagnosis and monitoring, leading to better patient care.

EXERCISE

- Explain the role of biomarkers in the diagnosis and management of pancreatic cancer.
- Describe the pathophysiological mechanisms of Inflammatory Bowel Disease (IBD).
- Discuss the clinical features and laboratory findings in acute pancreatitis.
- How is chronic pancreatitis diagnosed and managed?
- Explain the importance of fecal calprotectin in the assessment of inflammatory bowel disease.
- Describe the diagnostic approach to a patient with suspected pancreatic insufficiency.
- How do imaging techniques complement laboratory tests in diagnosing GI diseases?
- Discuss the role of endoscopic procedures in the management of gastrointestinal disorders.
- What are the common complications of untreated pancreatic diseases?
- Explain the impact of genetic factors on the development of gastrointestinal and pancreatic diseases.

CONCLUSION

The effective management of gastrointestinal and pancreatic diseases relies on a comprehensive understanding of their biochemical and physiological foundations. Advances in laboratory diagnostics and biomarker research have revolutionized the approach to these conditions, allowing for more precise and personalized treatment strategies. Continuous research and integration of clinical and laboratory data are essential for improving patient outcomes in GI and pancreatic diseases.

REFERENCES

[1] Sayegh, A.I. Basic notes in the physiology of the gastrointestinal tract. *JSM Gastroenterol Hepatol,* **2023**, *10*(3).

[2] Mcdowell, C.; Farooq, U.; Haseeb, M. Inflammatory Bowel Disease. [Updated 2023 Aug 4]. In: StatPearls [Internet]. Treasure Island (FL): StatPearls Publishing. **2025**. Available from: https://www.ncbi.nlm.nih.gov/books/NBK470312/

[3] Karpińska, M.; Czauderna, M. Pancreas—Its Functions, Disorders, and Physiological Impact on the Mammals' Organism. *Front. Physiol.,* **2022**, *13*, 807632. [http://dx.doi.org/10.3389/fphys.2022.807632] [PMID: 35431983]

[4] Dominguez-Munoz, J.E. Clinical pancreatology for practicing gastroenterologists. Juan Enrique Dominguez-Munoz (Editor), Wiley-Blackwell; **2021**.

CHAPTER 9

Hepatobiliary Disease

Abstract: Hepatobiliary diseases encompass a wide range of conditions affecting the liver, gallbladder, and bile ducts, with significant clinical implications. This chapter explores the biochemical and physiological aspects of the hepatobiliary system, providing insights into the pathophysiology, diagnosis, and management of common disorders such as hepatitis, cirrhosis, and biliary obstruction. Advances in laboratory medicine, including the use of specific biomarkers, have improved diagnostic accuracy and therapeutic strategies. Emphasis is placed on integrating laboratory results with clinical findings to enhance patient outcomes.

Keywords: Biliary, Biliary obstruction, Cirrhosis, Hepatic, Hepatitis.

LEARNING OBJECTIVES

After reading this chapter, one should be able to:

1. Be able to describe the physiological and biochemical functions of the hepatobiliary system and the pathophysiological mechanisms underlying common hepatobiliary diseases.
2. Be able to describe the causes, clinical signs, and symptoms of various disorders related to hepatobiliary function.
3. Be able to describe how laboratory investigations are important in the detection, diagnosis, and management.
4. Be able to describe specific laboratory investigations important to the study of disorders of hepatobiliary function.
5. Be able to recognise analytical methods available for their measurement.
6. Be able to select laboratory investigations and interpret the analytical results in the context of clinical signs and symptoms.

Simmi Kharb

KEY CONCEPTS: METABOLIC FUNCTIONS OF THE LIVER

Physiological and Biochemical Functions of Hepatobiliary System (Tables 1, 2) [1-4]

Table 1. Physiological and biochemical functions of the hepatobiliary system.

Function	Processes
Synthesis	**Protein Synthesis**: Albumin, clotting factors, various enzymes **Lipid Metabolism**: Lipoproteins, cholesterol, phospholipids **Carbohydrate Metabolism**: Glycogenesis, glycogenolysis, gluconeogenesis **Bile Production**: Emulsification and absorption of dietary fats
Conjugation	**Bilirubin Conjugation**: Conjugation of bilirubin with glucuronic acid for excretion in bile **Drug Metabolism**: Drugs and toxins become more water-soluble compounds **Hormone Metabolism**: Conjugation and inactivation of hormones (thyroid, steroid)
Detoxification	**Ammonia Detoxification**: Toxic ammonia converted to urea **Toxin and Drug Metabolism**: Detoxification of alcohol, drugs, and environmental toxins through Phase I and II xenobiotic metabolism

Table 2. Bilirubin.

Stage	Processes
Origin	**Heme Breakdown**: Breakdown of heme produces bilirubin **Formation of Unconjugated Bilirubin**: Heme converts into biliverdin and finally produces unconjugated bilirubin
Metabolism	**Transport to Liver**: Unconjugated bilirubin is transported bound to albumin **Conjugation in Liver**: Bilirubin is conjugated with glucuronic acid by UDP-glucuronosyltransferase
Transport and Excretion	**Biliary Excretion**: Conjugated bilirubin is excreted into bile and transported to the small intestine **Intestinal Metabolism**: Conjugated bilirubin is converted into urobilinogen by bacterial enzymes and excreted in feces as stercobilin **Urinary Excretion**: Some urobilinogen is reabsorbed; the rest is excreted in urine as urobilin

KEY CONCEPTS: ORIGIN, METABOLISM, AND TRANSPORT OF BILIRUBIN

Biochemical Basis, Relevant Lab Tests, Interpretation, and Applied Aspects

Biochemical Aspects of Hepatobiliary Disease

Biochemical Basis, Relevant Lab Tests, Interpretation and Applied Aspects of

Hepatobiliary Disease (Tables **3-5**; Figs. **1-4**) [1-4].

Table 3. Biochemical aspects of hepatobiliary disease.

Disease	Biochemical Aspects
Autoimmune Disease	**Primary Biliary Cholangitis (PBC):** Elevated Alkaline Phosphatase (ALP) and Antimitochondrial Antibodies (AMA) **Autoimmune Hepatitis (AIH):** Elevated transaminases (ALT, AST), hypergammaglobulinemia, positive Antinuclear Antibodies (ANA), and Anti-Smooth Muscle Antibodies (ASMA)
Cholestasis	**Biochemical Indicators:** Elevated ALP and Gamma-Glutamyl Transferase (GGT), increased conjugated bilirubin, decreased bile acid secretion, leading to jaundice and pruritus.
Drugs – Acute and Chronic	**Acute Drug-Induced Liver Injury (DILI):** Elevated ALT, AST, and sometimes ALP, with possible hyperbilirubinemia **Chronic DILI:** Persistent mild elevations of liver enzymes, potential progression to fibrosis or cirrhosis
Genetic (*e.g.*, A1AT Deficiency)	**Alpha-1 Antitrypsin (A1AT) Deficiency:** Low serum A1AT levels, accumulation of abnormal A1AT in hepatocytes, leading to elevated transaminases and risk of cirrhosis
Inflammatory and Infective Liver Disease (Hepatitis)	**Viral Hepatitis:** Elevated ALT, AST, and bilirubin levels, positive serological markers (HBsAg, anti-HCV) **Inflammatory Hepatitis:** Elevated inflammatory markers (CRP, ESR), and liver enzymes.
Liver Autoantibodies	**Autoimmune Hepatitis:** Positive ANA, ASMA, and liver/kidney microsomal antibodies (anti-LKM) **PBC:** Positive AMA
Liver Cirrhosis	**Biochemical Indicators:** Elevated bilirubin, prolonged Prothrombin Time (PT), low albumin, elevated liver enzymes (ALT, AST), and increased gamma-globulins.
Liver Transplantation	**Pre-Transplant:** Elevated liver enzymes, bilirubin, and markers of liver function (albumin, PT) **Post-Transplant:** Monitoring of liver enzymes, bilirubin, immunosuppressant levels, and signs of rejection or infection
Liver Tumors – Primary or Secondary	**Primary Liver Cancer (Hepatocellular Carcinoma):** Elevated Alpha-Fetoprotein (AFP), abnormal Liver Function Tests (LFTs), and imaging findings **Secondary (Metastatic) Liver Cancer:** Elevated liver enzymes, abnormal imaging, and tumor markers related to the primary cancer

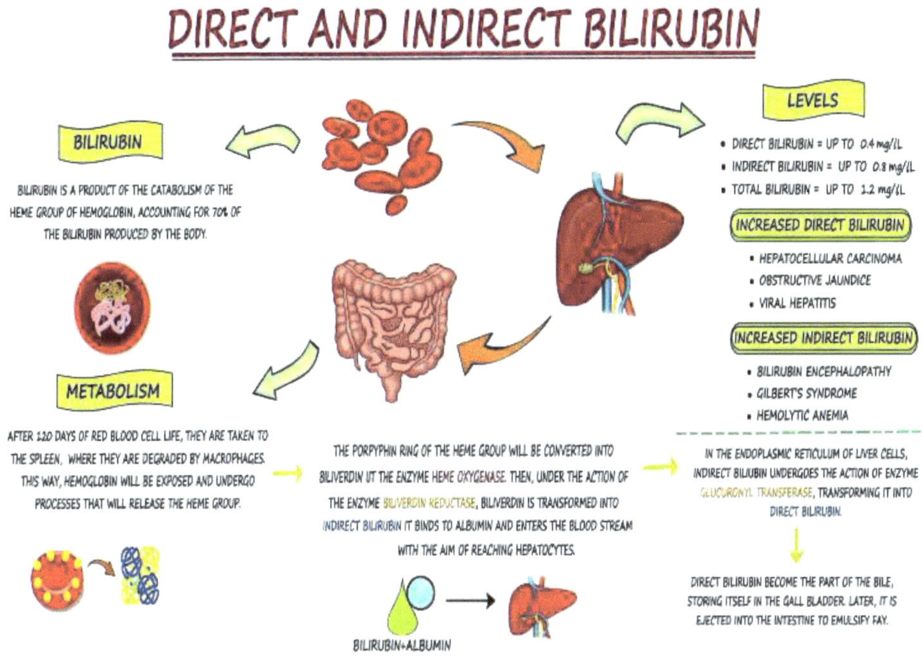

Fig. (1). Direct and indirect bilirubin.

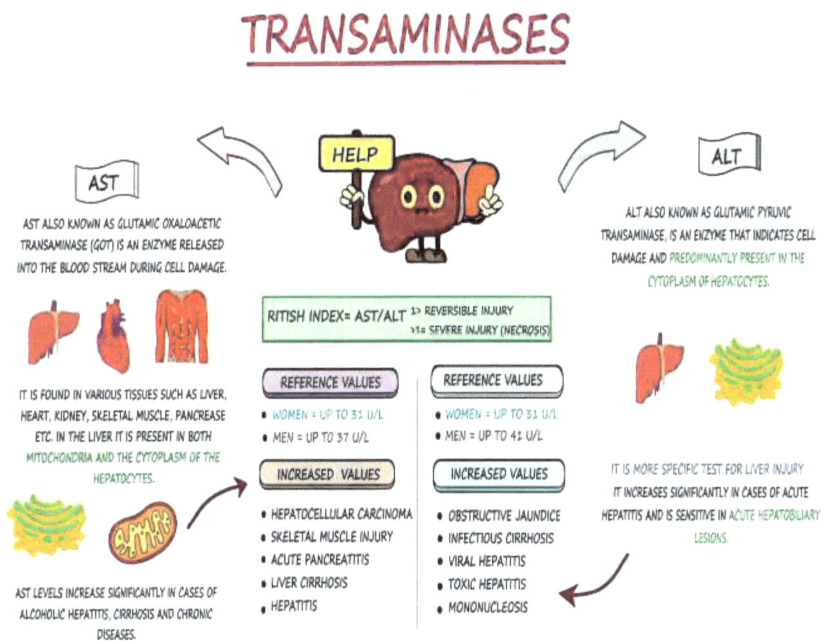

Fig. (2). Transaminases.

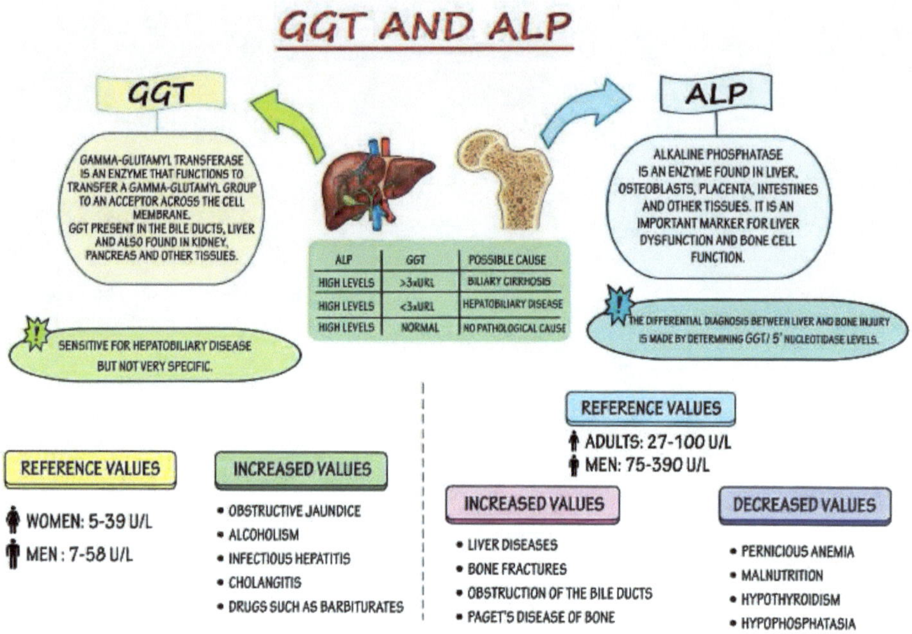

Fig. (3). GGT, ALP.

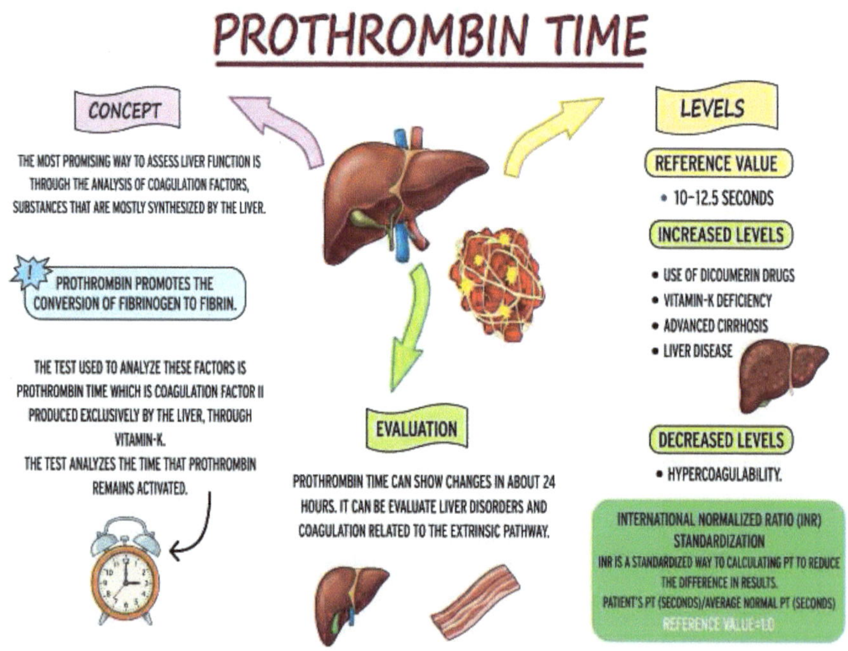

Fig. (4). Prothrombin time.

Biochemical basis, relevant lab tests, interpretation, and applied aspects

Biochemical basis of hepatobiliary disease

Table 4. Biochemical basis of hepatobiliary disease.

Liver Condition	Biochemical Basis
Acute Hepatitis	Elevated liver enzymes (ALT, AST), increased bilirubin levels (Fig. 1; 2)
Alcoholic Liver Disease	Elevated ALT, AST, and GGT. Increased MCV, elevated CDT (Fig. 3)
Allograft Rejection	Elevated liver enzymes (ALT, AST). Decreased bilirubin levels initially, potentially rising later
Biliary Damage and Dysfunction	Elevated ALP and GGT. Increased bilirubin levels (direct and indirect)
Chronic Hepatitis	Persistent elevation of ALT and AST. Viral load quantification (HCV RNA, HBV DNA)
Cirrhosis (Including Primary Biliary Cirrhosis)	Progressive liver enzyme elevation, decreased albumin, prolonged PT/INR. Elevated bilirubin levels
Elevated Bilirubin Levels	Increased total bilirubin, direct (conjugated) and indirect (unconjugated) bilirubin
Elevated Neonatal Bilirubin Levels	Increased total and direct bilirubin due to hepatic immaturity
Inherited Disorders of Bilirubin Metabolism	Variable unconjugated bilirubin levels due to genetic defects (*e.g.*, Gilbert syndrome)
Infectious Hepatitis	Elevated ALT and AST. Increased bilirubin levels. Viral load quantification (HCV RNA, HBV DNA)
Acetaminophen Overdose/Toxicity	Elevated ALT and AST due to hepatocellular damage. Elevated bilirubin and PT/INR in severe cases
Wilson's Disease	Decreased serum ceruloplasmin, increased hepatic copper. Elevated ALT, AST, and low serum copper levels

Table 5. Biochemical basis of hepatobiliary disease.

Condition	Relevant Laboratory Tests
Acute Hepatitis	- Hepatitis A, B, C serology (HBsAg, anti-HAV IgM, anti-HCV) Liver enzymes (ALT, AST) Total bilirubin and direct bilirubin levels (Fig. 1)
Alcoholic Liver Disease	- Liver enzymes (ALT, AST) - Gamma-Glutamyl Transferase (GGT), Mean Corpuscular Volume (MCV) - Carbohydrate-Deficient Transferrin (CDT)

(Table 5) cont.....

Condition	Relevant Laboratory Tests
Allograft Rejection	- Liver function tests (ALT, AST, bilirubin) - Immunosuppressant drug levels (*e.g.*, tacrolimus, cyclosporine) - Biopsy for histopathological evaluation
Biliary Damage and Dysfunction	- Liver enzymes (ALT, AST, GGT) (Fig. **1** - **3**) - Alkaline Phosphatase (ALP) and bilirubin levels - Imaging studies (MRCP, ERCP)
Chronic Hepatitis	- Hepatitis B, C serology (HBsAg, anti-HCV) - Liver enzymes (ALT, AST) - Viral load quantification (HCV RNA, HBV DNA)
Cirrhosis (Including Primary Biliary Cirrhosis)	- Liver function tests (ALT, AST, bilirubin, albumin, PT/INR) - Autoimmune markers (AMA for PBC) - Imaging (ultrasound, CT, MRI)
Elevated Bilirubin Levels (Conjugated, Unconjugated, Total)	- Total bilirubin, direct bilirubin, indirect bilirubin levels - Liver enzymes (ALT, AST) - Additional tests based on suspected cause (*e.g.*, genetic testing for inherited disorders) (Fig. **1**)
Elevated Neonatal Bilirubin Levels	- Total and direct bilirubin levels - Phototherapy level monitoring - Blood group compatibility testing for exchange transfusion (Fig. **1**; **4**)
Inherited Disorders of Bilirubin Metabolism	- Genetic testing for mutations (*e.g.*, UGT1A1 gene for Gilbert syndrome) - Liver function tests (ALT, AST) - Bilirubin levels and fractionation
Infectious Hepatitis	- Hepatitis A, B, C serology (HBsAg, anti-HAV IgM, anti-HCV) - Liver enzymes (ALT, AST) - Viral load quantification (HCV RNA, HBV DNA)
Acetaminophen Overdose/Toxicity	- Serum acetaminophen level (especially within 4-24 hours post-ingestion) - Liver enzymes (ALT, AST) - Prothrombin Time (PT), International Normalized Ratio (INR)
Wilson's Disease	- Serum ceruloplasmin, copper levels - 24-hour urinary copper excretion - Liver function tests (ALT, AST)

Biochemical basis, relevant lab tests, interpretation, and applied aspects: liver conditions

Relevant laboratory tests, analytical methods available

Analytical Methods Available for their Measurement (Table **6**).

Table 6. Analytical methods available.

Laboratory Test	Clinical Utility	Interpretation	Analytical Methods
Albumin (blood and ascites)	Assessing liver synthetic function and nutritional status	Decreased levels indicate liver dysfunction or malnutrition	Colorimetric assays
Alpha-1 Antitrypsin concentration and phenotype/genotype	Screening for alpha-1 antitrypsin deficiency	Low levels or abnormal phenotype/genotype indicate a deficiency	Electrophoresis, nephelometry
Alpha Fetoprotein (AFP)	Screening for Hepatocellular Carcinoma (HCC) and monitoring treatment response	Elevated levels suggest HCC or other liver tumors	Immunoassays
Ammonia	Monitoring hepatic encephalopathy and liver function	Elevated levels indicate impaired liver detoxification	Enzymatic assays
Bile Acids	Assessing liver function, particularly in cholestatic liver diseases	Elevated levels suggest impaired bile flow	Enzymatic methods, HPLC
Bilirubin – Conjugated, Unconjugated, Total	Assessing liver function and diagnosing jaundice	Elevated conjugated bilirubin indicates hepatobiliary disease	Diazo method for total bilirubin; enzymatic methods for conjugated bilirubin (Fig. **1**)
Transcutaneous Bilirubin Measurement	Screening for hyperbilirubinemia in neonates	Correlates with serum bilirubin levels; helps in determining the need for phototherapy	Spectrophotometry
Ceruloplasmin	Screening for Wilson's disease	Decreased levels suggest a copper metabolism disorder	Immunoturbidimetric assay
Ethanol	Detecting recent alcohol ingestion or evaluating alcohol-related liver disease	Presence indicates recent alcohol intake	Enzymatic assays
Ferritin	Assessing iron stores and diagnosing iron overload conditions	Elevated levels suggest iron overload or inflammation	Immunoassays
Hepatitis Serology	Diagnosing viral hepatitis (*e.g.*, Hepatitis A, B, C)	The presence of specific antibodies or antigens indicates current or past infection	Immunoassays
Liver Enzymes: ALT, AST, ALP, GGT	Assessing liver injury and function	Elevated levels indicate hepatocellular damage (ALT, AST), cholestasis (ALP, GGT)	Enzymatic assays

(Table 6) cont.....

Laboratory Test	Clinical Utility	Interpretation	Analytical Methods
Measurement of Immunosuppressant Drugs	Monitoring drug levels in transplant patients	Ensures therapeutic levels for immunosuppression without toxicity	Immunoassays, mass spectrometry
Acetaminophen	Assessing toxicity in overdose situations	Elevated levels indicate acetaminophen toxicity	Enzymatic assays

SUMMARY

Hepatobiliary diseases, such as hepatitis, cirrhosis, and biliary obstruction, pose complex diagnostic and therapeutic challenges. A profound understanding of their biochemical and physiological foundations is essential. Diagnosis typically entails a comprehensive fusion of clinical assessment and laboratory investigations, such as liver function tests and imaging studies. Management approaches vary based on the specific condition, encompassing pharmacological treatments, lifestyle adjustments, and surgical interventions. Progress in biomarker research has significantly bolstered the ability to diagnose and monitor these diseases, yielding enhanced patient care.

EXERCISES

1. Discuss the role of biomarkers in the diagnosis and management of hepatocellular carcinoma.
2. Describe the pathophysiological mechanisms of alcoholic liver disease.
3. Discuss the clinical features and laboratory findings in acute viral hepatitis.
4. How is non-alcoholic fatty liver disease diagnosed and managed?
5. Explain the importance of liver function tests in the assessment of hepatobiliary diseases.
6. Describe the diagnostic approach to a patient with suspected biliary obstruction.
7. How do imaging techniques complement laboratory tests in diagnosing hepatobiliary diseases?
8. Tabulate the role of lab investigations in the management of biliary disorders.
9. What are the common complications of untreated cirrhosis and their lab diagnosis?
10. Explain the impact of genetic factors on the development of hepatobiliary diseases.

CONCLUSION

The management of hepatobiliary diseases requires a deep understanding of their biochemical and physiological basis. Advances in laboratory diagnostics and biomarker research have led to more personalized treatment strategies.

Continuous research and integration of clinical and laboratory data are crucial for improving patient outcomes in hepatobiliary diseases.

REFERENCES

[1] Sherlock, S.; Dooley, J. *Diseases of the Liver and Biliary System,* 13th ed; Wiley-Blackwell, **2018**.

[2] Friedman, S.; Blumberg, R.S.; Saltzman, J.R. *Greenberger's CURRENT Diagnosis & Treatment Gastroenterology, Hepatology, & Endoscopy,* 4th ed; Lange, **2022**.

[3] Jameson, J.; Fauci, A.S.; Kasper, D.L.; Hauser, S.L.; Longo, D.L.; Loscalzo, J. *Harrison's Principles of Internal Medicine (22nd Ed.)*; McGraw-Hill Education, **2026**.

[4] Lonardo, A.; Ballestri, S. Perspectives of nonalcoholic fatty liver disease research: A personal point of view. *Exploration of Medicine,* **2020**, *1*(3), 85-107. [http://dx.doi.org/10.37349/emed.2020.00007]

CHAPTER 10

Immunology

Abstract: Immunology stands as a dynamic and intricate field with a pivotal role in understanding and addressing a diverse array of diseases. This chapter presents an in-depth exploration of the essential principles of immunology, encompassing both innate and adaptive immune responses. Furthermore, it delves into the intricacies of immune tolerance and autoimmunity. The ongoing evolution of immunological assays has significantly enhanced the diagnosis and management of immunological disorders, furnishing indispensable insights into patient care. The integration of immunological data with clinical findings is imperatively emphasized to bolster therapeutic strategies and improve patient outcomes.

LEARNING OBJECTIVES

After reading the chapter, students should be able to:

- Describe the principles, components, and control of the immune system.
- Describe the causes, clinical signs, and symptoms of disorders of the immune system.
- Describe how laboratory investigations are important in the detection, diagnosis, and management.
- Describe specific laboratory investigations important in their detection, diagnosis, and management.
- Recognize analytical methods available for their measurement.
- Select laboratory investigations and interpret the analytical results in the context of clinical signs and symptoms.

KEY CONCEPTS: BASICS, COMPONENTS, AND CONTROL OF THE IMMUNE SYSTEM

Principles, Components and Control of the Immune System (Tables 1-4) [1 - 4]

Simmi Kharb

Table 1. Components of the immune system.

Concept	Description
Immune System	A defense system against pathogens, comprising various organs, cells, and molecules
Innate Immunity	A non-specific defense system present from birth It includes physical barriers and immune cells (*e.g.*, macrophages and neutrophils)
Adaptive Immunity	Specific defense mechanisms that develop throughout life It involves T and B lymphocytes and provides long-term immunity
Antigens	Molecules (often from pathogens) that trigger an immune response by binding to specific antibodies or T-cell receptors
Antibodies	Proteins produced by B cells that neutralize pathogens or mark them for destruction by other immune cells
T Cells	Lymphocytes that mature in the thymus They play a role in directly killing infected cells (cytotoxic T cells) or coordinating immune responses (helper T cells)
B Cells	Lymphocytes that mature in bone marrow and produce antibodies specific to antigens encountered
Major Histocompatibility Complex (MHC)	Proteins that display antigens on cell surfaces for recognition by T cells These are crucial for immune response coordination
Cytokines	Signaling molecules secreted by immune cells that regulate inflammation, immune cell proliferation, and responses to infection
Immune response	A coordinated series of steps by the immune system to eliminate pathogens and restore homeostasis
Immune tolerance	Mechanisms that prevent the immune system from attacking the body's own cells and tissues
Immunological memory	The ability of the immune system to remember previous encounters with specific antigens, enabling faster and more effective responses upon re-exposure

KEY CONCEPTS: COMPONENTS OF IMMUNE SYSTEM: CELLS, LYMPHOID TISSUE, SOLUBLE COMPONENTS, AND MEDIATORS

Table 2. Components and mediators of the immune system.

Component	Description	Applied Aspects
Cells	Lymphocytes (T cells, B cells), macrophages, dendritic cells, neutrophils, and natural killer cells	-**Lymphocytes**: Adaptive immunity, memory response -**Macrophages**: Phagocytosis, antigen presentation -**Dendritic Cells**: Antigen presentation, activation of T cells -**Neutrophils**: Phagocytosis, inflammation -**Natural Killer Cells**: Attack virus-infected and tumor cells
Lymphoid Tissue	Lymph nodes, spleen, tonsils, and Mucosa-Associated Lymphoid Tissue (MALT)	-**Lymph Nodes**: Filter lymph, activate immune response -**Spleen**: Filters blood, immune responses to blood-borne pathogens -**MALT**: Protects mucosal surfaces, initial immune response
Soluble Components	Antibodies (immunoglobulins), complement proteins, cytokines (interleukins, interferons), and chemokines.	-**Antibodies**: Neutralization, opsonization, complement activation -**Complement Proteins**: Membrane attack complex, inflammation -**Cytokines**: Cell signaling, immune response modulation -**Chemokines**: Cell migration, inflammation
Mediators	Histamine, prostaglandins, leukotrienes, and nitric oxide that regulate inflammation, vascular permeability, and other immune responses	-**Histamine, Prostaglandins, Leukotrienes**: Vasodilation, inflammation -**Nitric Oxide**: Antimicrobial activity, vasodilation

KEY CONCEPTS: IMMUNE CELLS

Table 3. Components of the immune system: cells.

Concept	Description	Applied Aspects
Innate Immune System	Non-specific defense mechanisms that provide immediate protection against pathogens, including physical barriers and immune cells like macrophages and neutrophils.	-**Infection Control**: Recognizing and responding to pathogens quickly. -**Inflammation Regulation**: Understanding innate immune responses in diseases like sepsis

(Table 3) cont.....

Concept	Description	Applied Aspects
Lymphocytes	White blood cells involved in adaptive and innate immunity, including B and T cells.	-**Immunotherapy**: Engineering lymphocytes for targeted cancer treatments -**Autoimmune Diseases**: Understanding lymphocyte dysregulation in disorders like multiple sclerosis
B- Lymphocytes (B cells)	Part of adaptive immunity; they produce antibodies specific to antigens encountered	-**Vaccination**: Generating antibodies for long-term immunity against pathogens -**Diagnostic Testing**: Using B-cell responses for disease detection and monitoring
T- Lymphocytes (T cells)	Mature in the thymus and play roles in cell-mediated immunity, including cytotoxic T cells and helper T cells that regulate immune responses.	-**Transplantation**: Matching T cell receptors to prevent graft rejection -**Cancer Immunotherapy**: Engineering T cells to target and destroy cancer cells

KEY CONCEPTS: IMMUNOGLOBULIN PRODUCTION, STRUCTURE, AND FUNCTION

Table 4. Components of the immune system: immunoglobulin.

Concept	Description	Applied Aspects
Immunoglobulins (Antibodies)	Produced by B cells in response to antigens, crucial for immune defense	-**Specificity**: Recognizing and binding to specific antigens for neutralization or tagging for destruction
Structure	Y-shaped molecules consisting of *two heavy and two light chains linked by disulfide bonds* Each antibody has variable (antigen-binding) and constant regions	-**Engineering**: Modifying antibody structures for therapeutic purposes (*e.g.*, monoclonal antibodies)
Types (IgG, IgM, IgA, IgE)	Different classes with distinct roles: IgG (main in blood, crosses placenta), IgM (early response), IgA (mucosal immunity), IgE (allergic reactions)	-**Clinical Diagnostics**: Measuring antibody levels to diagnose infections or immune disorders
Production	Initiated by B-cell activation and differentiation into plasma cells that secrete antibodies	-**Vaccination**: Stimulating antibody production for disease prevention -**Therapeutics**: Producing antibodies for passive immunity treatments
Functions	-**Neutralization**: Blocking pathogen activity -**Opsonization**: Marking pathogens for phagocytosis -**Complement Activation**: Initiating immune responses	-**Autoimmune Diseases**: Targeting antibodies to reduce immune system attack on self-tissues -**Cancer Therapy**: Using antibodies to target cancer cells for destruction

(Table 4) cont.....

Concept	Description	Applied Aspects
Memory Response	Generation of memory B cells that facilitate rapid antibody production upon subsequent antigen exposure	-**Long-term Immunity**: Enhancing protection against recurring infections

KEY CONCEPTS: LABORATORY INVESTIGATIONS IN DETECTION, DIAGNOSIS AND MANAGEMENT (TABLE 5.1,5.2; 6,-11)

Table 5.1. Acute phase response.

Concept	Description	Applied Aspects
Acute Phase Response	Immediate systemic response to infection or tissue injury, involving changes in plasma protein levels (*e.g.*, C-reactive protein, cytokines).	-**Diagnostic Markers**: Monitoring acute inflammation or infection through biomarkers -**Therapeutic Targets**: Modulating acute phase proteins for disease management

Table 5.2. Adaptive immune system.

Concept	Description	Applied Aspects
Adaptive Immune System	Specific response involving T and B lymphocytes that target specific antigens, leading to immunological memory and long-term protection	-**Vaccination**: Inducing adaptive immune responses for disease prevention -**Immunotherapy**: Using antibodies or engineered T cells to treat diseases
Cellular Immunity	Involves T lymphocytes (*e.g.*, cytotoxic T cells, helper T cells) that directly attack infected or abnormal cells	-**Autoimmune Diseases**: Understanding T cell dysregulation in autoimmune conditions -**Transplantation**: Matching MHC for T cell recognition in organ transplantation
Humoral Immunity	Mediated by B lymphocytes and antibodies that neutralize pathogens or mark them for destruction by other immune cells	-**Antibody Therapeutics**: Using monoclonal antibodies to target specific diseases -**Diagnostic Assays**: Detecting antigen-specific antibodies in serological tests
Inflammatory Reactions	Non-specific response to injury or infection characterized by redness, heat, swelling, and pain, mediated by cytokines, chemokines, and immune cells.	-**Anti-inflammatory Drugs**: Modulating inflammatory responses for therapeutic purposes -**Wound Healing**: Understanding immune modulation in tissue repair processes.

KEY CONCEPTS: PRINCIPLES OF CELLULAR IMMUNITY, HUMORAL IMMUNITY AND INFLAMMATORY REACTIONS

Table 6. Principles of immunity reactions.

Concept	Description	Applied Aspects
Cellular Immunity	Immune response Involves T lymphocytes that recognize and destroy infected or abnormal cells. Includes cytotoxic T cells (CD8+) and helper T cells (CD4+)	-**Cancer Therapy**: Engineering T cells (*e.g.*, CAR-T cells) to target and kill cancer cells -**Transplantation**: Matching donor and recipient T cell receptors to prevent rejection
Humoral Immunity	Immune response mediated by B lymphocytes that produce antibodies specific to antigens. Involves the production of memory B cells for long-term immunity	-**Vaccination**: Inducing antibody production for protection against infections -**Passive Immunity**: Using monoclonal antibodies for immediate protection against diseases
Inflammatory Reactions	The body's non-specific response to injury or infection, characterized by redness, heat, swelling, and pain Mediated by immune cells, cytokines, and other molecules	-**Anti-inflammatory Drugs**: Developing medications to reduce inflammation in conditions like arthritis -**Wound Healing**: Understanding the role of inflammation in tissue repair and regeneration

KEY CONCEPTS: COMPLEMENT SYSTEM

Table 7. Principles of immunity: complement system.

Concept	Description	Applied Aspects
Complement System	A group of proteins that enhance immune responses, including opsonization, inflammation, and cell lysis	-**Immune Defense**: Enhancing pathogen clearance through opsonization and Membrane Attack Complex (MAC) formation
Alternative Pathway	Activated spontaneously by microbial surfaces lacking complement inhibitors, leading to C3 convertase formation	-**Microbial Defense**: Complementing antibody-mediated immunity by targeting pathogens directly
Classical Pathway	Triggered by antigen-antibody complexes binding to C1q, initiating a cascade leading to C3 convertase formation	-**Immune Complex Diseases**: Understanding complement involvement in autoimmune diseases like lupus

(Table 7) cont.....

Concept	Description	Applied Aspects
Mannose-binding Lectin (MBL) Pathway	Initiated by MBL binding to mannose residues on pathogen surfaces, activating lectin-associated serine proteases (MASPs) and leading to C3 convertase formation	-**Innate Immunity**: Enhancing immune responses against pathogens not targeted by antibody-mediated pathways
Opsonization	Coating of pathogens with complement proteins (*e.g.*, C3b) to enhance phagocytosis by immune cells	-**Vaccine Adjuvants**: Using complement activation to improve immune responses in vaccines
Inflammation	Mediated by complement activation products like C3a and C5a, which induce vasodilation, chemotaxis, and activation of immune cells	-**Therapeutic Targeting**: Modulating complement-mediated inflammation in diseases like rheumatoid arthritis
Membrane Attack Complex (MAC)	Formation of MAC (C5b-C9) on microbial membranes, leading to cell lysis and destruction	-**Infectious Diseases**: Understanding the complement's role in defense against bacterial infections

KEY CONCEPTS: *CYTOKINES:*

Table 8. Principles of immunity: cytokines.

Concept	Description	Applied Aspects
Cytokines	Small proteins secreted by various cells that regulate immunity, inflammation, and hematopoiesis	-**Immune Modulation**: Targeting cytokines for therapeutic interventions in autoimmune diseases and cancer
Colony Stimulating Factors (CSFs)	Stimulate the production and differentiation of hematopoietic stem cells into specific blood cell types (*e.g.*, granulocytes, macrophages)	-**Hematopoietic Recovery**: Using CSFs to accelerate recovery after chemotherapy or bone marrow transplantation
Hematopoietic Factors	Regulate the proliferation and differentiation of blood cell precursors in bone marrow	-**Stem Cell Therapy**: Enhancing efficacy of stem cell transplantation through hematopoietic factors
Interferons	Proteins that inhibit viral replication and modulate immune responses against viruses and tumors	-**Antiviral Therapies**: Using interferons to treat viral infections like hepatitis and multiple sclerosis
Interleukins	Cytokines that mediate communication between leukocytes, regulating immune responses and inflammation	-**Immunotherapy**: Utilizing interleukins to enhance immune responses against cancer cells -**Autoimmune Diseases**: Targeting interleukin pathways for treatment

(Table 8) cont.....

Concept	Description	Applied Aspects
Tumor Necrosis Factors (TNFs)	Involved in inflammation, apoptosis, and immune regulation, crucial in autoimmune diseases and cancer	-**Anti-TNF Therapies**: Blocking TNF to treat conditions like rheumatoid arthritis and Crohn's disease -**Cancer Treatment**: Targeting TNF pathways in tumor therapy

KEY CONCEPTS: *HYPERSENSITIVITY REACTION TYPES I – IV (OR V):*

Table 9. Principles of immunity: hypersensitivity reactions.

Type	Description	Applied Aspects
Type I: Immediate Hypersensitivity	IgE-mediated reaction triggered by allergens, leading to mast cell and basophil activation and release of histamine and other mediators	-**Allergy Management**: Identifying allergens and administering antihistamines or epinephrine (in severe cases) -**Desensitization Therapy**: Gradual exposure to reduce allergic responses
Type II: Cytotoxic Hypersensitivity	Antibody-mediated reaction causing destruction of host cells by complement activation or Antibody-Dependent Cell-mediated Cytotoxicity (ADCC)	-**Blood Transfusion**: Matching blood types to prevent hemolytic reactions -**Autoimmune Hemolytic Anemia**: Understanding antibody-mediated destruction of red blood cells
Type III: Immune Complex-Mediated	Formation of immune complexes (antigen-antibody complexes) that deposit in tissues, triggering inflammation and tissue damage	-**Autoimmune Diseases**: Managing conditions like lupus, where immune complexes deposit in organs -**Serum Sickness**: Recognizing hypersensitivity to therapeutic proteins or drugs
Type IV: Delayed-Type Hypersensitivity	Mediated by T cells and macrophages, leading to inflammatory reactions hours to days after exposure to antigens (*e.g.*, contact dermatitis, tuberculosis skin test)	-**Contact Dermatitis**: Identifying and avoiding allergens causing skin reactions -**Diagnostic Testing**: Assessing immune responses in tuberculosis and other infections
Type V: Stimulatory Hypersensitivity	Involves autoantibodies that stimulate receptors or interfere with normal cellular signaling, leading to abnormal cell function	-**Graves' Disease**: Understanding autoantibodies that stimulate thyroid hormone production

KEY CONCEPTS: IMMUNOGLOBULINS

Table 10. Principles of immunity: immunoglobulin.

Concept	Description	Applied Aspects
Function	Antibodies produced by B cells that recognize and neutralize specific antigens, promoting immune responses like opsonization and neutralization.	-**Therapeutics**: Using monoclonal antibodies for targeted therapy in diseases like cancer.
Gene Rearrangement	Process in B cells in which segments of antibody genes (V, D, J segments) recombine to create diverse antigen-binding sites.	-**Immune Disorders**: Studying gene rearrangement errors in autoimmune diseases and immunodeficiencies.
Structure	Y-shaped proteins with two heavy and two light chains linked by disulfide bonds, with variable and constant regions.	-**Vaccine Design**: Understanding antibody structure for antigen selection in vaccine development.

KEY CONCEPTS: INNATE IMMUNE SYSTEM

Table 11. Principles of immunity: innate immune system.

Concept	Description	Applied Aspects
Innate Immune System	Non-specific defense mechanism that provides immediate protection against pathogens, including physical barriers (skin, mucous membranes), immune cells (macrophages, neutrophils), and molecular components (complement system)	-**Infection Control**: Quickly recognizing and responding to pathogens -**Inflammation Regulation**: Understanding and managing innate immune responses in diseases such as sepsis
Lymphocytes	White blood cells involved in adaptive and innate immunity, including B cells, T cells, and Natural Killer (NK) cells	-**Immunotherapy**: Engineering lymphocytes for targeted cancer treatments -**Autoimmune Diseases**: Understanding lymphocyte dysregulation in conditions such as multiple sclerosis
B Lymphocytes (B cells)	Part of adaptive immunity that produces antibodies specific to antigens encountered Mature in the bone marrow and differentiate into plasma cells that secrete antibodies and memory B cells	-**Vaccination**: Generating antibodies for long-term immunity against pathogens -**Diagnostic Testing**: Using B-cell responses for disease detection and monitoring

(Table 11) cont.....

Concept	Description	Applied Aspects
T Lymphocytes (T cells)	Mature in the thymus that play a role in cell-mediated immunity, including cytotoxic T cells (CD8+) that directly kill infected or abnormal cells, and helper T cells (CD4+) that regulate immune responses	-**Transplantation**: Matching donor and recipient T cell receptors to prevent graft rejection -**Cancer Immunotherapy**: Engineering T cells to target and destroy cancer cells

BIOCHEMICAL BASIS, RELEVANT LAB TESTS, INTERPRETATION, AND APPLIED ASPECTS

Laboratory Investigations and Interpretation of Analytical Results in the Context of Clinical Signs and Symptoms (Tables 12,13.1, 13.2) [1 - 4]

Table 12. Laboratory investigations and interpretation of analytical results.

Condition	Biochemical Basis	Relevant Lab Tests	Interpretation of Results	Applied Aspects
Allergy and Anaphylaxis	IgE-mediated hypersensitivity reactions	- Serum IgE levels - Skin prick tests - Specific IgE antibody tests	- Elevated IgE levels indicate allergic sensitization - Positive skin tests confirm specific allergens	- Allergy management with antihistamines and epinephrine - Desensitization therapies
Autoimmune Diseases	Autoantibodies target self-antigens	- ANA (Antinuclear Antibody) - RF (Rheumatoid Factor) - Anti-dsDNA - Anti-CCP	- Positive ANA in lupus - Elevated RF in rheumatoid arthritis - Anti-dsDNA in lupus	-Immuno-suppressive therapy - Biological agents targeting specific pathways
Endocrine	Hormonal imbalances due to immune-mediated gland dysfunction	- Thyroid function tests (TSH, T3, T4) - Cortisol levels - Autoantibodies (*e.g.*, anti-TPO, anti-TG)	- Abnormal TSH levels in thyroid disorders - Positive anti-TPO in Hashimoto's thyroiditis	- Hormone replacement therapy - Immune modulation
Gastrointestinal (GI) Tract	Immune-mediated inflammation of the GI tract	- Fecal calprotectin - CRP (C-reactive protein) - Specific autoantibodies (*e.g.*, anti-tTG for celiac disease)	- Elevated fecal calprotectin in inflammatory bowel disease - Positive anti-tTG in celiac disease	- Dietary management for celiac disease - Immunosuppressive therapy for IBD

(Table 12) cont.....

Condition	Biochemical Basis	Relevant Lab Tests	Interpretation of Results	Applied Aspects
Liver	Autoimmune hepatitis, cirrhosis, or other liver conditions	- Liver function tests (ALT, AST, ALP, bilirubin) - Autoantibodies (*e.g.*, ANA, SMA)	- Elevated liver enzymes indicate liver damage - Positive autoantibodies suggest autoimmune hepatitis	- Immuno-suppressive therapy for autoimmune hepatitis - Liver transplantation in severe cases
Renal	Immune-mediated glomerulonephritis and other kidney conditions	- Serum creatinine - Urinalysis - Autoantibodies (*e.g.*, ANCA, anti-GBM)	- Elevated creatinine indicates kidney dysfunction - Positive ANCA in vasculitis	- Immunosuppressive therapy - Dialysis or kidney transplantation in advanced cases
Rheumatic and Articular	Autoimmune inflammation of joints and connective tissues	- RF - Anti-CCP - CRP, ESR	- Elevated RF and anti-CCP in rheumatoid arthritis - High CRP/ESR indicate inflammation	- NSAIDs, DMARDs, and biological agents for treatment
Skin	Autoimmune or allergic skin conditions	- Skin biopsy - Serum IgE - Autoantibodies (*e.g.*, anti-BP180 in bullous pemphigoid)	- Biopsy confirms specific skin conditions - Elevated IgE in atopic dermatitis	- Topical and systemic immunosuppressive therapy - Allergen avoidance
Lymphoid Malignancy	Uncontrolled proliferation of lymphoid cells	- Flow cytometry – Immuno-phenotyping - Genetic testing (*e.g.*, FISH, PCR)	- Abnormal cell populations in flow cytometry - Genetic abnormalities confirm specific malignancies	- Chemotherapy, radiotherapy, and targeted therapy - Stem cell transplantation
B Cell	Abnormal B-cell function or proliferation	- Serum immunoglobulin levels - Flow cytometry - Bone marrow biopsy	- Low immunoglobulin levels in B cell deficiency - Abnormal B cell populations in malignancies	- Immuno-globulin replacement therapy - Targeted therapies for B cell malignancies
T Cell	Abnormal T-cell function or proliferation	- Flow cytometry - T Cell Receptor (TCR) analysis - Lymphocyte proliferation assays	- Reduced T cell counts in immunodeficiency - Abnormal TCR patterns in malignancies	- Antiretroviral therapy for HIV/AIDS - Immuno-therapy for T cell malignancies

(Table 12) cont.....

Condition	Biochemical Basis	Relevant Lab Tests	Interpretation of Results	Applied Aspects
Primary Immune Deficiency	Genetic defects affecting immune function	- Serum immunoglobulin levels - Flow cytometry - Genetic testing	- Low immunoglobulin levels indicate antibody deficiency - Genetic mutations confirm specific immunodeficiencies	- Immuno-globulin replacement - Hematopoietic stem cell transplantation
Secondary Immune Deficiency	Acquired conditions affecting immune function	- CD4 count (HIV) - Serum immunoglobulin levels - Nutritional assessments	- Low CD4 count in HIV/AIDS - Low immunoglobulins in secondary immunodeficiency	- Antiretroviral therapy for HIV - Addressing underlying causes (*e.g.*, malnutrition, cancer treatment)
Immunoglobulin Overproduction	Excessive production of monoclonal antibodies or cryoglobulins	- Serum Protein Electrophoresis (SPEP) - Immuno-Fixation Electrophoresis (IFE) - Cryoglobulin test	- Monoclonal spike in SPEP indicates gammopathy - Positive cryoglobulin test indicates cryoglobulinemia	- Targeted therapies for multiple myeloma - Plasmapheresis for cryoglobulinemia
Immunologic Impact of Transplantation	Immune responses to transplanted organs or tissues	- HLA typing -Cross-matching - Immuno-suppressive drug levels	- Matching HLA reduces rejection risk - Therapeutic drug monitoring ensures effective immuno-suppression	- Immuno-suppressive therapy to prevent rejection - Monitoring for infections and complications post-transplant

Relevant laboratory tests for detection, diagnosis, management, and monitoring

Table 13.1. Laboratory investigations and analytical methods.

Laboratory Test	Analytical Methods
Albumin (blood and ascites)	Bromocresol green dye-binding method, immunoassay
Alpha-1 antitrypsin concentration and phenotype/genotype	Nephelometry, immunoassay, and isoelectric focusing
Alpha Fetoprotein (AFP)	Immunoassay and chemiluminescent immunoassay
Ammonia	Enzymatic assay and ion-selective electrode
Bile acids	Enzymatic cycling method and High-Performance Liquid Chromatography (HPLC)

(Table 13.1) cont.....

Laboratory Test	Analytical Methods
Bilirubin – conjugated, unconjugated, total	Diazo method, Jendrassik-Grof method, and HPLC
Transcutaneous bilirubin measurement	Transcutaneous bilirubinometry
Ceruloplasmin	Immunoassay and nephelometry
Ethanol	Enzymatic assay and Gas Chromatography (GC)
Ferritin	Immunoassay and chemiluminescent immunoassay
Hepatitis serology	Enzyme-Linked Immunosorbent Assay (ELISA), chemiluminescent immunoassay
Iron	Colorimetric assay and Atomic Absorption Spectrometry (AAS)
Liver enzymes: ALT, AST, ALP, GGT	Kinetic UV assay and enzymatic colorimetric assay
Measurement of immunosuppressant drugs	Liquid Chromatography-Tandem Mass Spectrometry (LC-MS/MS) and immunoassay
Acetaminophen	Enzymatic assay, immunoassay, and HPLC
PT-INR	Coagulometry and prothrombin time assay
Autoantibodies: Liver Kidney Microsomal Ab, smooth muscle antibody, *etc.*	Immunoassay and indirect immunofluorescence

Relevant laboratory tests, analytics methods available, and applied aspects

Table 13.2. Laboratory investigations and analytical methods.

Test	Detection/Diagnosis/Management	Analytical Methods	Applied Aspects
Autoantibodies			
-Skin Basement Membrane, Intercellular Cement	Diagnosis of bullous pemphigoid and pemphigus vulgaris	- Indirect immunofluorescence - ELISA	- Confirm diagnosis of blistering skin diseases
-Nuclear (ANA)	Screening for systemic autoimmune diseases (*e.g.*, SLE)	- Indirect immunofluorescence - ELISA	- High sensitivity for detecting Systemic Lupus Erythematosus (SLE)
-Smooth Muscle	Diagnosis of autoimmune hepatitis	- Indirect immunofluorescence - ELISA	- Confirm autoimmune hepatitis
-Mitochondrial	Diagnosis of primary biliary cirrhosis	- Indirect immunofluorescence - ELISA	- Specific for primary biliary cirrhosis
-Neutrophil Cytoplasmic (ANCA)	Diagnosis of vasculitis (*e.g.*, granulomatosis with polyangiitis)	- Indirect immunofluorescence - ELISA	- Identifies and differentiates types of vasculitides

(Table 13.2) cont.....

Test	Detection/Diagnosis/Management	Analytical Methods	Applied Aspects
-Glomerular Basement Membrane	Diagnosis of Goodpasture's syndrome	- Indirect immunofluorescence - ELISA	- Detects autoantibodies specific to renal and pulmonary basement membranes
-Rheumatoid Factor	Diagnosis of rheumatoid arthritis	- Latex agglutination - Nephelometry - ELISA	- Indicates rheumatoid arthritis, but can be present in other conditions
-Cyclic Citrullinated Peptides (Anti-CCP)	Diagnosis of rheumatoid arthritis	- ELISA	- High specificity for rheumatoid arthritis
-Double Stranded DNA (Anti-dsDNA)	Diagnosis and monitoring of SLE	- ELISA - Immuno-fluorescence	- Specific marker for SLE activity
-Extractable Nuclear Antigens (ENA)	Diagnosis of various autoimmune diseases (*e.g.*, Sjogren's syndrome, SLE)	- ELISA - Immunoblotting	- Differentiates between various autoimmune diseases
-Thyroid (*e.g.*, anti-TPO, anti-TG)	Diagnosis of autoimmune thyroid diseases (*e.g.*, Hashimoto's thyroiditis)	- ELISA - Chemiluminescent assays	- Confirms autoimmune thyroid disease
-Pancreas (*e.g.*, anti-GAD, anti-IA2)	Diagnosis of type 1 diabetes	- ELISA - Radio-immunoassay	- Detects autoantibodies in type 1 diabetes
-Adrenal (*e.g.*, anti-21-hydroxylase)	Diagnosis of autoimmune Addison's disease	- ELISA - Immunoblotting	- Confirms autoimmune adrenal insufficiency
-Ovary, Testis (*e.g.*, anti-ovarian, anti-testicular)	Diagnosis of autoimmune ovarian and testicular failure	- ELISA - Immuno-fluorescence	- Detects autoantibodies causing infertility
-Tissue Transglutaminase (tTG), Endomysial	Diagnosis of celiac disease	- ELISA - Indirect immunofluorescence	- High specificity and sensitivity for celiac disease
-Intrinsic Factor	Diagnosis of pernicious anemia	- ELISA - Immunoassay	- Confirms diagnosis of pernicious anemia
Complete Blood Count (CBC)	Overall health status, infection, anemia, and immune status	- Automated cell counters	- Routine monitoring for various conditions
HLA Typing	Organ transplantation compatibility and disease association	- PCR - Flow cytometry	- Ensures donor -recipient compatibility in transplantation; identifies genetic predispositions

(Table 13.2) cont.....

Test	Detection/Diagnosis/Management	Analytical Methods	Applied Aspects
Immunoglobulin Quantification and IgG Subclasses	Assessment of immune function, diagnosis of immunodeficiencies	- Nephelometry - ELISA	- Diagnoses primary and secondary immunodeficiencies, monitors immune status
Lymphocyte Subsets (CD3, 4, 8, 16/56, 19)	Immune profiling in various diseases (*e.g.*, HIV, lymphomas)	- Flow cytometry	- Monitors immune reconstitution in HIV; evaluates lymphocyte populations in cancers
Monoclonal Protein Identification and Quantification in Serum, CSF, and Urine	Diagnose and monitor monoclonal gammopathies (*e.g.*, multiple myeloma)	- Serum Protein Electrophoresis (SPEP) - Immunofixation Electrophoresis (IFE)	- Detects and quantifies monoclonal proteins
Evaluation of Electrophoretic and Immunofixation Results in Serum, Urine, and CSF	Detects monoclonal and oligoclonal gammopathies	- Electrophoresis - Immunofixation	- Identifies and monitors multiple myeloma, Waldenström macroglobulinemia, and other plasma cell disorders
Total and Specific IgE	Diagnose allergic conditions and monitor response to allergens	- ELISA - RAST (radio-allergosorbent test)	- Identifies specific allergen sensitivities, monitors allergic disease management
Tryptase	Diagnosis of mast cell activation disorders (*e.g.*, anaphylaxis, mastocytosis)	- ELISA - Immunoassay	- Elevated levels indicate mast cell degranulation, useful in anaphylaxis diagnosis

SUMMARY

Immunology is the study of the immune system, which protects the body against infections and diseases. It includes innate and adaptive immunity. Disorders can lead to autoimmunity, hypersensitivities, and immunodeficiencies. Diagnostic tools like immunological assays are essential. Treatment often involves tailored immunomodulatory therapies. Clinical and immunological data integration is crucial for effective management and improved patient outcomes.

EXERCISES

- What are the differences between innate and adaptive immunity?

- Explain the roles of T cells and B cells in the immune response.
- Discuss how immune tolerance works and how its breakdown can cause autoimmunity.
- How are immunological tests used to diagnose autoimmune diseases?
- What is the clinical importance of flow cytometry in immunology?
- Describe the process of antigen presentation and why it is important in the immune response.
- How do monoclonal antibodies function as immunotherapeutic agents?
- Discuss the roles of cytokines in the immune response and their potential for treatment.
- What are the common immunodeficiency disorders, and how are they diagnosed?
- Explain the impact of hypersensitivity reactions on health and disease management.

CONCLUSION

It is crucial to understand the principles of immunology and the mechanisms of immune-related diseases for accurate diagnosis and treatment. The advancements in immunological assays have greatly improved our capacity to address these conditions effectively. Ongoing research and the convergence of clinical and immunological data are imperative for elevating patient outcomes in the field of immunology.

REFERENCES

[1] Parham, P. *The Immune System,* 5th international student edition; Garland Science, Taylor & Francis Group, LLC: New York, NY, **2021**.

[2] Abbas, A.K.; Lichtman, A.H.; Pillai, S. *Basic Immunology: Functions and Disorders of the Immune System,* 7th ed; Elsevier Health Sciences, **2023**.

[3] Medzhitov, R.; Iwasaki, A. Exploring new perspectives in immunology *Cell,* **2024**, *187*, 2079-2094. [http://dx.doi.org/10.1016/j.cell.2024.03.038]

[4] Murphy, K.; Weaver, C. *Janeway's Immunobiology,* 10th International Student Edition; New York, NY : W.W. Norton and Company, **2022**.

CHAPTER 11

Inborn Errors of Metabolism

Abstract: Inborn Errors of Metabolism (IEMs) are a diverse group of genetic disorders resulting from defects in metabolic pathways. These disorders can lead to the accumulation of toxic substances or the deficiency of critical compounds, manifesting in a wide range of clinical symptoms. This chapter overviews the biochemical mechanisms underlying IEMs, highlights key diagnostic laboratory tests, and examines therapeutic approaches.

LEARNING OBJECTIVES

After reading this chapter, one should be able to:

- Understand the biochemical basis and pathophysiology of common Inborn Errors of Metabolism (IEMs).
- Identify the clinical manifestations and diagnostic criteria for IEMs.
- Evaluate the role of laboratory tests in the diagnosis, monitoring, and management of IEMs.
- Explore current and emerging therapeutic strategies for managing IEMs.
- Discuss the genetic basis and inheritance patterns of IEMs.

KEY CONCEPTS: *CHALLENGES WITH NEWBORN SCREENING.*

Biochemical Basis and Pathophysiology of Common Inborn Errors of Metabolism (IEMs) (Table 1) [1 - 4]

Simmi Kharb

Table 1. Biochemical basis and pathophysiology of common inborn errors of metabolism.

Key Concept	Description
Rationale for Screening	Early identification of metabolic disorders to prevent severe complications through timely intervention.
Overview of the Process	Newborn screening involves blood tests, typically using a heel prick, to detect specific metabolic disorders. Early diagnosis facilitates early treatment, reducing morbidity and mortality.
Challenges with Newborn Screening	Technical Challenges: Ensuring the accuracy and reliability of tests Logistical Challenges: Implementing efficient sample collection and processing Ethical and Legal Challenges: Addressing consent, privacy, and potential for false positives/negatives Includes false positives/negatives, limited availability of tests for all IEMs, variability in screening protocols, and the need for confirmatory testing.
Diseases Appropriate for Newborn Screening	Characteristics: - Prevalence: Sufficiently common to justify screening - Seriousness: Causes significant morbidity/mortality if untreated - Treatability: Effective treatment or intervention available - Detectability: Reliable, timely, and cost-effective test available.
Newborn Screening Process	Steps Involved: - Sample Collection: Blood sample typically collected *via* heel prick - Laboratory Analysis: Samples analyzed using various biochemical and genetic tests - Result Reporting: Timely reporting of results to healthcare providers and families - Follow-Up: Immediate follow-up for positive or ambiguous results to confirm diagnosis and initiate treatment.

KEY CONCEPTS: *DISEASES APPROPRIATE FOR NEWBORN SCREENING – CHARACTERISTICS* (TABLE 2)

Table 2. Clinical manifestations and diagnostic criteria for IEMs.

Key Concept	Applied Aspects
Rationale for Screening and Overview	- Early Detection: Identifying congenital disorders early to prevent severe health problems and improve outcomes - Cost-Effectiveness: Reducing long-term healthcare costs by preventing complications - Public Health Impact: Improving overall public health by addressing common congenital disorders early.

Key Concept	Applied Aspects
Challenges with Newborn Screening	- Technical Challenges: Ensuring the accuracy and reliability of screening tests; maintaining quality control - Logistical Challenges: Efficient sample collection, handling, and processing; timely reporting of results - Ethical and Legal Challenges: Addressing issues related to informed consent, data privacy, and potential psychological impact on families.
Diseases Appropriate for Newborn Screening	- Prevalence: Disorders must be sufficiently common to justify population-wide screening - Seriousness: Conditions that cause significant morbidity and mortality if untreated - Treatability: Availability of effective treatment or intervention that can significantly alter the course of the disease - Detectability: Reliable, timely, and cost-effective screening tests must be available.
Newborn Screening Process	- Sample Collection: Typically involves collecting a blood sample *via* heel prick - Laboratory Analysis: Use of biochemical and genetic tests such as tandem mass spectrometry, immunoassays, and molecular techniques - Result Reporting: Timely communication of results to healthcare providers and families - Follow-Up: Immediate follow-up for positive or ambiguous results to confirm diagnosis and initiate treatment.

BIOCHEMICAL BASIS, RELEVANT LAB TESTS, INTERPRETATION, AND APPLIED ASPECTS

Role of Laboratory Tests in Diagnosis, Monitoring, and Management of IEMs (Tables 3,4) [1 - 4]

Table 3. Role of laboratory tests in diagnosis, monitoring, and management of IEMs.

Condition	Biochemical Basis	Relevant Lab Tests	Interpretation of Results	Applied Aspects
Congenital Hypothyroidism	Deficiency in thyroid hormone production due to thyroid gland dysgenesis or dyshormonogenesis.	TSH, Free T4, and Total T4	Elevated TSH and low Free T4/Total T4	Early treatment with thyroid hormone replacement to prevent intellectual disability.

(Table 3) cont.....

Condition	Biochemical Basis	Relevant Lab Tests	Interpretation of Results	Applied Aspects
Cystic Fibrosis	A mutation in the CFTR gene, leading to thick and sticky mucus production.	Sweat chloride test, CFTR gene mutation analysis, and Immunoreactive Trypsinogen (IRT)	Elevated sweat chloride levels, presence of CFTR mutations, and high IRT in newborns	Management includes airway clearance techniques, pancreatic enzyme supplementation, and addressing infections.
Enzyme Deficiencies: Biotinidase, Galactokinase	Biotinidase: Deficiency in biotin recycling. Galactokinase: Deficiency in converting galactose to galactose-1-phosphate.	Biotinidase activity assay, Galactokinase enzyme activity, and Galactose-1-phosphate uridyltransferase (GALT)	Low biotinidase activity, elevated blood/urine galactose, and low GALT activity	Biotin supplementation for biotinidase deficiency; avoidance of galactose for galactokinase deficiency.
Fatty Acid Oxidation Disorders	Defects in enzymes responsible for β-oxidation of fatty acids.	Acylcarnitine profile, blood glucose, and urinary organic acids	Abnormal acylcarnitine profile, hypoglycemia, and elevated urinary organic acids	Avoidance of fasting, dietary management, and carnitine supplementation.
Hemoglobinopathies	Structural abnormalities in hemoglobin (*e.g.*, Sickle Cell, Thalassemia).	Hemoglobin electrophoresis and DNA analysis	Abnormal hemoglobin patterns and identification of mutations	Regular monitoring, transfusions, chelation therapy for iron overload, and hydroxyurea for sickle cell disease.
Homocystinuria	Deficiency in cystathionine β-synthase leading to elevated homocysteine.	Plasma amino acids and Homocysteine levels	Elevated methionine and homocysteine	Vitamin B6, B12, folate supplementation, and dietary methionine restriction.
Lysosomal Storage Diseases	Deficiency in specific lysosomal enzymes (*e.g.*, lipidoses, hexosaminidases, Fabry's disease).	Enzyme activity assays, Genetic testing, and Urinary Glycosaminoglycans (GAGs)	Reduced enzyme activity, presence of genetic mutations, and elevated urinary GAGs	Enzyme replacement therapy, substrate reduction therapy, and symptom management.

(Table 3) cont.....

Condition	Biochemical Basis	Relevant Lab Tests	Interpretation of Results	Applied Aspects
Maple Syrup Urine Disease (MSUD)	Deficiency in branched-chain α-keto acid dehydrogenase complex.	Plasma amino acids and urinary organic acids	Elevated leucine, isoleucine, valine; presence of branched-chain α-keto acids in urine	Dietary restriction of branched-chain amino acids, and thiamine supplementation.
Phenylketonuria (PKU)	Deficiency in phenylalanine hydroxylase leading to elevated phenylalanine.	Plasma phenylalanine levels and genetic testing	Elevated plasma phenylalanine	Dietary restriction of phenylalanine and tyrosine supplementation.
Tyrosinemia	Deficiency in Fumarylacetoacetate Hydrolase (FAH) leading to elevated tyrosine.	Plasma amino acids and urinary succinylacetone	Elevated tyrosine and presence of succinylacetone in urine	**Tyrosinemia**

Relevant laboratory tests, analytics methods available, and applied aspects

Table 4. Role of laboratory tests in diagnosis, analytical methods, and interpretation.

Test	Analytical Method	Applied Aspects	Interpretation
Acylcarnitines	Tandem Mass Spectrometry (MS/MS)	Diagnosis of fatty acid oxidation disorders and organic acidemias	Abnormal acylcarnitine profiles indicate specific metabolic disorders
Alpha-1-antitrypsin genotyping	Polymerase Chain Reaction (PCR), sequencing	Identifying genetic variants associated with alpha-1-antitrypsin deficiency	Detection of PiZ and PiS alleles associated with increased risk of emphysema and liver disease
Amino acids	High-Performance Liquid Chromatography (HPLC), MS/MS	Diagnosis of amino acid metabolism disorders (*e.g.*, PKU, MSUD)	Elevated or decreased levels of specific amino acids in plasma or urine
Hemoglobin electrophoresis	Electrophoresis, HPLC	Diagnosis of hemoglobinopathies (*e.g.*, sickle cell disease, thalassemia)	Identification of abnormal hemoglobin variants
Immunoreactive trypsinogen	Immunoassay	Newborn screening for cystic fibrosis	Elevated levels suggest CF; further genetic testing is required

(Table 4) cont.....

Test	Analytical Method	Applied Aspects	Interpretation
Organic acids	Gas Chromatography-Mass Spectrometry (GC-MS)	Diagnosis of organic acidemias	Elevated levels of specific organic acids in urine indicate metabolic disorders
Specific enzyme testing (*e.g.*, biotinidase, galactokinase)	Enzyme activity assays	Diagnosis of specific enzyme deficiencies	Low enzyme activity levels indicate a deficiency and a need for treatment
Sweat chloride and conductivity	Sweat test (quantitative pilocarpine iontophoresis)	Diagnosis of cystic fibrosis	Elevated sweat chloride levels confirm a CF diagnosis

SUMMARY

IEMs are caused by genetic mutations that lead to dysfunctional enzymes or transport proteins in metabolic pathways. This disruption can result in various clinical presentations, from acute metabolic crises to chronic developmental delays. The chapter covers the pathophysiology of several common IEMs, including amino acidopathies, organic acidemias, and fatty acid oxidation disorders. Diagnostic strategies involve biochemical assays, genetic testing, and newborn screening programs. Management includes dietary modifications, enzyme replacement therapies, and emerging gene therapies.

EXERCISES

- Explain the biochemical basis of Phenylketonuria (PKU) and its clinical manifestations.
- Describe the diagnostic laboratory tests used to confirm a diagnosis of Maple Syrup Urine Disease (MSUD).
- Discuss the therapeutic approaches for managing fatty acid oxidation disorders.
- How does genetic testing complement biochemical assays in the diagnosis of IEMs?
- Evaluate gene therapy's potential in treating IEMs and discuss current challenges.

CONCLUSION

Understanding the complex nature of IEMs requires a multidisciplinary approach that combines biochemical, genetic, and clinical expertise. Early diagnosis through newborn screening and targeted genetic testing is crucial for effective management. Advances in therapeutic strategies, including gene therapy, hold promise for improving the prognosis of individuals with IEMs. Ongoing research and clinical trials are essential for developing innovative treatments and improving patient care.

REFERENCES

[1] Blau, N.; Duran, M.; Gibson, K.M.; Vici, C.D., Eds. *Physician's guide to the diagnosis, treatment, and follow-up of inherited metabolic diseases*; Berlin, Heidelberg Springer **2023**. [http://dx.doi.org/10.1007/978-3-642-40337-8]

[2] den Hollander, B, Hoytema van Konijnenburg EMM, Hewitson B The Metabolic Treatabolome and Inborn Errors of Metabolism Knowledgebase therapy tool: Do not miss the opportunity to treat! *J Inherit Metab Dis,* **2025**, *48*(1), e12835. [http://dx.doi.org/10.1002/jimd.12835]

[3] Rodrigues, KF, Yong WTL, Bhuiyan MSA, Siddiquee S, Shah MD, Venmathi Maran BA., Ed. Current understanding on the genetic basis of key metabolic disorders: a review. *Biology (Basel),* **2022**, *11*(9), 1308. [http://dx.doi.org/10.3390/biology11091308]

[4] Ijaz, A.; Abbas, S.; Shabbir, M.; Badshah, Y.; Abid, F.; Afsar, T.; Razak, S. Inherited metabolic disorders: presentation, clinical types, laboratory diagnosis and genetic markers. *Orphanet J Rare Dis,* **2025**, *20*(1), 422. [http://dx.doi.org/10.1186/s13023-025-03979-8]

CHAPTER 12

Infectious Diseases

Abstract: This chapter emphasises the practical applications of infectious diseases, highlighting the crucial role of laboratory diagnostics in detecting, managing, and monitoring infections. It addresses various types of infections, such as viral, bacterial, and parasitic diseases, providing in-depth information on the biochemical foundations, pertinent laboratory tests, and result interpretation. The goal is to offer clinical chemists the necessary knowledge to make meaningful contributions to patient care and public health efforts.

LEARNING OBJECTIVES

After going through the chapter, one should be able to:

- Describe the role of laboratory testing in the diagnosis and monitoring of infectious diseases.
- Recommend and interpret laboratory tests for the diagnosis and management of infectious diseases.

KEY CONCEPTS *ROLE OF LABORATORY AND SEROLOGY TESTING IN DIAGNOSIS AND MONITORING OF INFECTIOUS DISEASE:*

Role of Laboratory Testing in Diagnosis and Monitoring of Infectious Diseases (Tables 1,2) [1, 2]

Simmi Kharb

Table 1. Role of laboratory testing in diagnosis and monitoring of infectious diseases.

Key Concept	Explanation	Applied Aspects	Clinical Relevance
Pathogenesis	Study of how infectious agents cause disease	Understanding microbial mechanisms and host responses	Aids in developing targeted treatments and vaccines
Microbial Identification	Techniques for identifying infectious agents (bacteria, viruses, fungi, parasites)	Culture, PCR, serology, and MALDI-TOF MS	Critical for accurate diagnosis and appropriate treatment
Antimicrobial Resistance	Mechanisms by which microbes resist treatment	Antibiotic stewardship and resistance gene detection	Prevents treatment failure and informs antibiotic policy
Diagnostic Testing	Methods for detecting infectious agents or their effects	Blood tests, imaging, and molecular diagnostics	Essential for confirming diagnoses and guiding treatment
Therapeutic Approaches	Treatment options for infectious diseases	Antibiotics, antivirals, antifungals, and supportive care	Enhances patient outcomes and reduces morbidity and mortality

DISEASE

Relevant Laboratory Tests for Infectious Diseases for Detection, Diagnosis, Management, or Monitoring

Table 2. Role of laboratory testing in diagnosis and monitoring of infectious diseases.

Aspect	Description
Biochemical Basis	The biochemical basis of infectious diseases involves understanding the interaction between pathogens (bacteria, viruses, fungi, parasites) and the host's cellular and molecular mechanisms. This includes the ability of a pathogen to invade host tissues, evade the immune system, and produce toxins or other virulence factors that cause disease
Relevant Laboratory Tests	
1. Blood Cultures	To detect bacterial or fungal pathogens in the bloodstream
2. Polymerase Chain Reaction (PCR)	To identify specific genetic material from pathogens Useful for detecting viruses and bacteria
3. Serology Tests	To measure antibodies or antigens in blood to detect infections, *e.g.*, HIV, hepatitis, and syphilis
4. Complete Blood Count (CBC)	To assess the overall health of the blood, indicating infections through changes in white blood cells
5. Urinalysis	To detect pathogens or abnormalities in urine, indicating urinary tract infections

(Table 2) cont.....

Aspect	Description
6. Imaging Studies (X-ray, CT, MRI)	Visualizes infections in organs and tissues
7. Culture and Sensitivity	Grows pathogens from samples (*e.g.*, sputum, wound swabs) to identify the organism and test antibiotic susceptibility
8. Rapid Antigen Tests	Quickly detects specific antigens from pathogens (*e.g.*, influenza, strep throat)
Interpretation of Results:	
Positive Blood Cultures	Indicates the presence of pathogens in the blood, suggesting sepsis or systemic infection
PCR Results	Positive results confirm the presence of the pathogen's genetic material, indicating an active infection
Serology Tests	Presence of specific antibodies indicates current or past infection; high levels of IgM suggest recent infection, while IgG suggests past exposure or immunity
CBC Results	Elevated white blood cell count (leukocytosis) typically indicates infection; specific patterns can suggest bacterial *vs.* viral infections
Urinalysis	Presence of bacteria, white blood cells, or nitrites indicates a urinary tract infection
Imaging Studies	Identifies structural abnormalities, abscesses, or inflammation associated with infections
Culture and Sensitivity	Identifies specific pathogens and guides appropriate antibiotic therapy based on susceptibility results
Rapid Antigen Tests	Positive results quickly confirm the presence of specific pathogens, allowing for timely treatment
Applied Aspects:	
1. Diagnosis	Accurate identification of the causative pathogen is crucial for effective treatment and management
2. Treatment	Laboratory results guide the choice of antimicrobial therapy, ensuring targeted and effective treatment
3. Public Health	Surveillance and monitoring of infectious diseases help in controlling outbreaks and implementing preventive measures
4. Infection Control	Identifying and isolating infectious patients prevents the spread of pathogens within healthcare settings
5. Prognosis and Monitoring	Regular monitoring of infection markers and response to treatment guides clinical decisions and improves patient outcomes

RELEVANT LABORATORY TESTS FOR DETECTION, DIAGNOSIS, MANAGEMENT, OR MONITORING AND ANALYTICS METHODS AVAILABLE

Recommendation and Interpretation of Laboratory Tests for the Diagnosis and Management of the Infectious Diseases (Tables 3,4) [1 - 5]

Table 3. Role of laboratory testing in diagnosis, monitoring, and analytical methods available.

Test	Detection/Diagnosis/Monitoring	Analytical Methods	Applied Aspects
Blood Cultures	Detection of bacteremia or fungemia	Blood sample inoculated into culture media and incubated to detect microbial growth	Identifies bloodstream infections, guides antimicrobial therapy
Polymerase Chain Reaction (PCR)	Detection of specific pathogens (*e.g.*, viruses, bacteria)	Amplification of pathogen-specific DNA or RNA sequences	High sensitivity and specificity, rapid results, useful for early diagnosis
Serology Tests	Diagnosis of infections like HIV, hepatitis, and syphilis	Enzyme-Linked Immunosorbent Assay (ELISA), Western blot	Detects antibodies or antigens, monitors immune response, determines past or current infection
Complete Blood Count (CBC)	General assessment of infection	Automated hematology analyzers	Indicates infection through elevated WBCs, differentiates between bacterial and viral infections
Urinalysis	Detection of urinary tract infections	Dipstick tests, microscopic examination, and urine culture	Identifies the presence of bacteria, white blood cells, and nitrites in urine
Imaging Studies (X-ray, CT, MRI)	Visualization of infections in organs and tissues	Radiography, computed tomography, and magnetic resonance imaging	Identifies structural abnormalities, abscesses, and inflammation
Culture and Sensitivity	Identification of pathogens and antibiotic susceptibility	Growth of pathogens on selective media, antibiotic susceptibility testing (*e.g.*, disk diffusion, MIC)	Determines specific pathogens and appropriate antibiotic treatment

(Table 3) cont.....

Test	Detection/Diagnosis/Monitoring	Analytical Methods	Applied Aspects
Rapid Antigen Tests	Quick detection of specific pathogens (*e.g.*, influenza, strep throat)	Lateral flow assays, immunochromatographic assays	Provides rapid results and is useful for point-of care testing
Enzyme Immunoassays (EIA)	Detection of bacterial toxins (*e.g.*, *C. difficile*) or viral antigens	ELISA and chemiluminescent immunoassays	Quantitative measurement of antigens or toxins; supports diagnosis and monitoring
Hematology Tests	Assessment of bloodborne infections	Automated blood analyzers and flow cytometry	Measures various blood parameters, indicates the presence and type of infection
Biochemical Markers	Monitoring infection and inflammation (*e.g.*, CRP, procalcitonin)	Immunoassays and turbidimetric assays	Helps in monitoring infection severity and treatment response
Molecular Diagnostics	Detection of genetic material from pathogens (*e.g.*, TB, HIV)	Real-time PCR, Next-Generation Sequencing (NGS)	Highly sensitive and specific, useful for early detection and epidemiological studies
Next-Generation Sequencing (NGS)	Comprehensive pathogen detection and resistance profiling	High-throughput DNA sequencing	Identifies multiple pathogens and resistance genes in a single test, useful for outbreak investigation
Immunofluorescence Assays	Detection of viral antigens (*e.g.*, respiratory viruses)	Direct and indirect immunofluorescence	Visualizes pathogen-specific antigens, supports rapid diagnosis

RELEVANT LABORATORY TESTS FOR DETECTION, DIAGNOSIS, AND MANAGEMENT, OR MONITORING ANALYTICS METHODS AVAILABLE AND APPLIED TO ASPECTS OF INFECTIOUS DISEASES

Table 4. Role of laboratory testing in diagnosis, analytical methods, and interpretation.

Disease	Clinical Signs and Symptoms	Relevant Laboratory Tests	Analytical Methods	Applied Aspects
HIV	Fever, weight loss, swollen lymph nodes, and opportunistic infections	HIV Antibody/Antigen (4th Gen), HIV RNA PCR, and CD4 Count	ELISA, Western Blot, and PCR	Early detection, monitoring of viral load and immune function, and guidance for antiretroviral therapy
Hepatitis A	Jaundice, fatigue, abdominal pain, and loss of appetite	Anti-HAV IgM and Anti-HAV IgG	ELISA and Chemiluminescent Immunoassays	Diagnosis of acute and past infections, as well as he assessment of immune status
Hepatitis B	Jaundice, fatigue, dark urine, and joint pain	HBsAg, Anti-HBs, Anti-HBc, and HBV DNA	ELISA, PCR, and Quantitative assays	Determination of infection status, monitoring treatment, and assessing immunity
Hepatitis C	Fatigue, jaundice, abdominal pain, and dark urine	Anti-HCV, HCV RNA PCR, and HCV Genotyping	ELISA, PCR, and Sequencing	Diagnosis of acute/chronic infection, and guiding antiviral therapy
Hepatitis E	Jaundice, fatigue, abdominal pain, and nausea	Anti-HEV IgM, Anti-HEV IgG, and HEV RNA PCR	ELISA and PCR	Diagnosis of acute infection and monitoring outbreaks
Herpes Simplex Virus 1/2	Oral/genital lesions, fever, and body aches	HSV-1/2 IgG/IgM, HSV PCR, and Viral Culture	ELISA, PCR, and Cell Culture	Diagnosis of primary and recurrent infections and guiding antiviral therapy
EBV (Epstein-Barr Virus)	Sore throat, fever, swollen lymph nodes, and fatigue	EBV VCA IgM/IgG, EBNA IgG, and EBV DNA PCR	ELISA and PCR	Diagnosis of primary and past infections and monitoring immunocompromised patients

(Table 4) cont.....

Disease	Clinical Signs and Symptoms	Relevant Laboratory Tests	Analytical Methods	Applied Aspects
CMV (Cytomegalovirus)	Fever, sore throat, swollen glands, and fatigue	CMV IgM/IgG and CMV DNA PCR	ELISA and PCR	Diagnosis of active and past infections, monitoring congenital infections, and immunocompromised patients
M. Tuberculosis	Chronic cough, weight loss, night sweats, and hemoptysis	Tuberculin Skin Test (TST), Interferon-Gamma Release Assays (IGRAs), Sputum AFB Stain, and TB PCR	Mantoux Test, ELISA, Ziehl-Neelsen Stain, and PCR	Detection of latent and active infections and guiding treatment
SARS-CoV-2	Fever, cough, shortness of breath, and loss of taste/smell	SARS-CoV-2 PCR, Rapid Antigen Tests, and Serology (IgM/IgG)	RT-PCR, Lateral Flow Assays, and ELISA	Confirming active infection, monitoring immunity and past infection, and public health surveillance
Syphilis	Chancre, rash, fever, and swollen lymph nodes	RPR, VDRL, FTA-ABS, TP-PA, and Treponemal EIA	Flocculation Tests, ELISA, and Particle Agglutination	Diagnosis of stage infection and monitoring treatment
Lyme Disease	Erythema migrans, fever, headache, and fatigue	ELISA for Borrelia burgdorferi antibodies, and Western Blot	ELISA and Western Blot	Diagnosis of early and late stages, and confirmatory testing

SUMMARY

Infectious diseases pose a significant challenge in clinical medicine. The early detection, accurate diagnosis, and effective management of these diseases heavily rely on laboratory diagnostics. This chapter discusses key infections, such as HIV, hepatitis, and tuberculosis, focusing on the biochemical markers and laboratory techniques used in their detection. It emphasizes the crucial need to understand the clinical implications of laboratory results and the necessity for continuous advancements in diagnostic methodologies to keep up with emerging pathogens.

EXERCISES

- Discuss the importance of the Polymerase Chain Reaction (PCR) in the diagnosis of infectious diseases.
- Describe the steps involved in the Enzyme-Linked Immunosorbent Assay (ELISA) and its applications in infectious disease diagnostics.

- Discuss the role of laboratory diagnostics in the management of HIV. Include the types of tests and their clinical significance.
- What are the challenges associated with the detection of Latent Tuberculosis Infection (LTBI)?
- Analyze the impact of Rapid Diagnostic Tests (RDTs) on the management of malaria in endemic regions.

CONCLUSION

Laboratory diagnostics are crucial in the battle against infectious diseases. The precise and prompt identification of pathogens, combined with a thorough understanding of the clinical context, is essential for effective patient management and outbreak control. With the emergence of new technologies, clinical chemists must remain informed and adaptable to ensure the best possible patient outcomes.

REFERENCES

[1] Baron, E.J. Clinical Microbiology Procedures Handbook. 5th edition, Wiley: **2023**.

[2] Balows, A. Manual of clinical microbiology 13th edition, Wiley; **2024**.

[3] Rapszky, G.A. Rapid molecular assays versus blood culture for bloodstream infections: a systematic review and meta-analysis *eClinicalMedicine,* **2025**, *79*, 103028. [http://dx.doi.org/10.1016/j.eclinm.2024.103028]

[4] Moyo, E.; Mhango, M.; Moyo, P.; Dzinamarira, T.; Chitungo, I.; Murewanhema, G. Emerging infectious disease outbreaks in Sub-Saharan Africa: Learning from the past and present to be better prepared for future outbreaks. *Front Public Health,* **2023**, *11*, 1049986. [http://dx.doi.org/10.3389/fpubh.2023.1049986]

[5] Zayed, R.A.; Omran, D.; Zayed, A.A. COVID-19 clinical and laboratory diagnosis overview. *J. Egypt. Public Health Assoc.,* **2021**, *96*(1), 25. [http://dx.doi.org/10.1186/s42506-021-00087-w] [PMID: 34406520]

CHAPTER 13

Iron and Hemoglobin Disorders, Including Porphyrias

Abstract: Understanding iron and hemoglobin disorders, such as porphyrias, is crucial for effective clinical diagnostics and management. This chapter provides in-depth insights into the biochemical foundations of these conditions, emphasizing the pivotal role of iron in the body, the genetic and molecular underpinnings of hemoglobinopathies, and the metabolic pathways associated with porphyrias. By examining pertinent laboratory tests and their interpretation, the chapter is designed to empower clinical chemists with the essential knowledge required for precise diagnosis and successful management of these disorders.

LEARNING OBJECTIVES

After reading through this chapter, one should be able to:

1. Understand Iron Metabolism: Gain insight into the regulation of iron absorption, transport, storage, and its role in hemoglobin synthesis.
2. Identify Hemoglobin Disorders: Learn about various hemoglobinopathies, including sickle cell disease and thalassemia.
3. Explore the Biochemistry of Porphyrias: Understand the biochemical basis of porphyrias and their clinical implications.
4. Interpret Diagnostic Tests: Develop skills in interpreting laboratory tests related to iron metabolism, hemoglobin disorders, and porphyrias.
5. Clinical Management: Understand the clinical management strategies for disorders of iron metabolism and hemoglobin, with a focus on therapeutic approaches.

Simmi Kharb

KEY CONCEPTS: IRON AND HEME METABOLISM AND THEIR REGULATION

Regulation of Iron Absorption, Transport, Storage, and its Role in Hemoglobin Synthesis (Tables 1-3.1, 3.2) [1-5]

Table 1. Iron absorption, transport, storage, and its role in hemoglobin synthesis.

Concept	Details
Iron Absorption	Iron is absorbed primarily in the duodenum. Non-heme iron (Fe^{3+}) is reduced to Fe^{2+} by cytb and then transported into enterocytes by DMT1. Heme iron is directly absorbed *via* Heme Carrier Protein 1 (HCP1).
Iron Transport	Once in the enterocyte, iron can be stored as ferritin or exported to the bloodstream by ferroportin. In the blood, Fe^{2+} is oxidized to Fe^{3+} by hephaestin and ceruloplasmin, and then bound to transferrin for transport (Fig. 1).
Iron Storage	Iron is stored in the liver, spleen, and bone marrow in the form of ferritin and hemosiderin.
Heme Synthesis	Heme is synthesized in the mitochondria and cytosol of cells, primarily in the bone marrow and liver. The process involves eight enzymatic steps, starting with the condensation of glycine and succinyl-CoA to form δ-aminolevulinic acid (ALA).
Heme Degradation	Heme is degraded to biliverdin by heme oxygenase, then reduced to bilirubin. Bilirubin is transported to the liver, conjugated with glucuronic acid, and excreted in bile.
Regulation of Iron Homeostasis	Hepcidin, produced by the liver, is the key regulator. It binds to ferroportin, causing its internalization and degradation, thus reducing iron absorption and release from macrophages.
Regulation of Heme Synthesis	Heme synthesis is regulated primarily at the level of ALA synthase, which is inhibited by heme (negative feedback).

KEY CONCEPTS: APPLIED ASPECTS OF IRON AND HEME METABOLISM

Table 2. Applied aspects of iron and heme metabolism.

Aspect	Details
Dietary Iron Intake	Iron can be obtained from dietary sources such as red meat (heme iron) and vegetables (non-heme iron). Vitamin C enhances non-heme iron absorption.
Iron Supplementation	Used to treat iron deficiency. Oral ferrous sulfate is common, but intravenous iron may be needed in severe cases.
Phlebotomy	Therapeutic phlebotomy is used to treat iron overload conditions such as hereditary hemochromatosis.

(Table 2) cont.....

Aspect	Details
Chelation Therapy	Deferoxamine and other iron chelators are used to treat iron overload by promoting excretion.
Monitoring Iron Levels	Serum ferritin, transferrin saturation, and Total Iron-Binding Capacity (TIBC) are used to assess iron status.

Table 3.1. Applied aspects of iron metabolism.

Disorder	Biochemical Basis
Iron Overload	Excessive iron absorption or multiple blood transfusions lead to iron accumulation in tissues, causing oxidative damage and organ dysfunction. Hepcidin deficiency is a common cause of hereditary hemochromatosis.
Iron Deficiency	Insufficient iron intake, increased requirements (*e.g.*, pregnancy), or chronic blood loss lead to reduced hemoglobin synthesis and microcytic hypochromic anemia.

KEY CONCEPTS: BIOCHEMICAL BASIS OF IRON OVERLOAD AND IRON DEFICIENCY

Implications of iron overload and iron deficiency

Table 3.2. Applied aspects of iron metabolism.

Disorder	Clinical Signs and Symptoms	Implications
Iron Overload	Fatigue, joint pain, abdominal pain, liver cirrhosis, diabetes, cardiomyopathy	Increased risk of liver cancer, diabetes, and heart disease. Requires regular monitoring and treatment to prevent organ damage.
Iron Deficiency	Fatigue, pallor, shortness of breath, dizziness, and pica	Reduced work capacity, impaired cognitive function, and increased risk of preterm birth in pregnant women. Requires dietary changes, supplementation, and treatment of underlying causes.

KEY CONCEPTS: APPLIED ASPECTS OF ENZYMATIC DEFECTS OF HEME SYNTHESIS (LEADING TO PORPHYRIAS)

Biochemistry of Porphyrias: Biochemical Basis of Porphyrias and their Clinical Implications (Tables 4-5) [3]

Table 4. Applied aspects of heme synthesis.

Aspect	Key Concepts	Applied Aspects
Heme Biosynthesis	- Heme is synthesized through a multi-step process involving eight enzymes - Begins with the condensation of glycine and succinyl-CoA to form δ-Aminolevulinic Acid (ALA). - Key steps: formation of Porphobilinogen (PBG), hydroxymethylbilane, uroporphyrinogen, coproporphyrinogen, protoporphyrinogen, and finally heme.	- Defects in these enzymes lead to various types of porphyrias - Symptoms depend on the specific enzyme affected and can include photosensitivity, abdominal pain, and neuropsychiatric symptoms - Management includes avoidance of triggers, symptomatic treatment, and in some cases, heme therapy.
Heme Metabolism	- Heme is broken down into biliverdin by heme oxygenase, then reduced to bilirubin. - Bilirubin is transported to the liver, conjugated with glucuronic acid, and excreted in bile.	- Disorders in heme metabolism can lead to jaundice and bilirubin accumulation - Management includes treating the underlying cause, supporting liver function, and sometimes phototherapy in neonatal jaundice.
Iron Absorption, Transport, and Storage	- Iron is absorbed primarily in the duodenum. - Transported in blood bound to transferrin. - Stored in liver, spleen, and bone marrow as ferritin and hemosiderin.	- Ensuring adequate iron intake and absorption is crucial for preventing iron deficiency anemia. - Management of iron overload conditions (*e.g.*, hereditary hemochromatosis) includes regular phlebotomy or iron chelation therapy.
Iron Deficiency	- Results from inadequate iron intake, increased need (*e.g.*, pregnancy), or chronic blood loss. - Leads to reduced hemoglobin synthesis and microcytic hypochromic anemia.	- Treatment includes dietary changes, oral iron supplements, or intravenous iron in severe cases - Addressing underlying causes, such as gastrointestinal bleeding, is also essential.
Iron Overload	- Excessive iron absorption or multiple blood transfusions lead to iron accumulation in tissues. - Hepcidin deficiency is a common cause of hereditary hemochromatosis.	- Requires regular monitoring of iron levels (serum ferritin, transferrin saturation). - Treatment includes therapeutic phlebotomy, iron chelation therapy, and lifestyle modifications to limit iron intake.

Table 5. Biochemical basis of porphyrias.

Type of Porphyria	Enzyme Deficiency	Accumulated Substrate	Clinical Features
Acute Intermittent Porphyria (AIP)	Porphobilinogen Deaminase (PBGD)	δ-Aminolevulinic Acid (ALA) and Porphobilinogen (PBG)	Abdominal pain, neuropsychiatric symptoms, and dark urine
Porphyria Cutanea Tarda (PCT)	Uroporphyrinogen Decarboxylase (UROD)	Uroporphyrin	Photosensitivity, skin fragility, and blistering
Erythropoietic Protoporphyria (EPP)	Ferrochelatase (FECH)	Protoporphyrin	Photosensitivity and liver dysfunction
Hereditary Coproporphyria (HCP)	Coproporphyrinogen Oxidase (CPOX)	Coproporphyrinogen III	Abdominal pain, neuropsychiatric symptoms, and photosensitivity
Variegate Porphyria (VP)	Protoporphyrinogen Oxidase (PPOX)	Protoporphyrinogen IX	Abdominal pain, neuropsychiatric symptoms, and photosensitivity
Congenital Erythropoietic Porphyria (CEP)	Uroporphyrinogen III Synthase (UROS)	Uroporphyrin I and Coproporphyrin I	Severe photosensitivity and hemolytic anemia

KEY CONCEPTS: TYPES OF PORPHYRIAS AND THEIR BIOCHEMICAL BASIS

Biochemical basis, relevant lab tests, interpretation, and applied aspects

Interpretation of Laboratory Tests Related to Iron Metabolism, Hemoglobin Disorders, and Porphyrias (Tables 6-8; Fig. 1) [1, 3, 5]

Table 6. Laboratory tests related to iron metabolism, hemoglobin disorders, and porphyrias, interpretation.

Condition	Biochemical Basis	Relevant Lab Tests	Interpretation of Results	Applied Aspects
Acute and Chronic Porphyrias	Deficiency in heme biosynthesis enzymes.	Urine, blood, stool porphyrin analysis, and enzyme assays.	Elevated ALA and PBG in urine (AIP); elevated fecal protoporphyrin (VP).	Avoidance of triggers and management of acute attacks with hemin.

(Table 6) cont.....

Condition	Biochemical Basis	Relevant Lab Tests	Interpretation of Results	Applied Aspects
Anemia Secondary to Malignancy	Impaired erythropoiesis due to cancer-induced cytokines.	CBC, serum ferritin, and erythropoietin levels.	Normocytic or microcytic anemia; elevated ferritin and low or normal erythropoietin.	Treat underlying malignancy, erythropoiesis-stimulating agents, and iron supplementation if indicated.
G6PD Deficiency	Deficiency in the G6PD enzyme, leading to oxidative stress in RBCs.	G6PD enzyme assay, hemoglobin levels, and reticulocyte count.	Reduced G6PD activity; hemolytic anemia during oxidative stress.	Avoid oxidative stressors and provide supportive care during hemolytic episodes.
Hemochromatosis	Mutations in the HFE gene, leading to increased iron absorption.	Serum ferritin, transferrin saturation, and genetic testing.	Elevated ferritin and transferrin saturation; positive genetic test.	Regular phlebotomy, iron chelation therapy, and monitoring of organ function.
Hemoglobinopathies	Genetic mutations in hemoglobin genes.	Hemoglobin electrophoresis, CBC, and genetic testing.	Abnormal hemoglobin variants, anemia, and specific RBC morphologies.	Regular monitoring of hemoglobin levels and managing complications.
Intravascular Hemolysis	Destruction of RBCs within blood vessels.	Haptoglobin levels, LDH, and hemoglobinuria.	Decreased haptoglobin, elevated LDH, and the presence of hemoglobin in urine.	Identifying and treating the underlying cause.
Iron Deficiency Anemia	Inadequate iron intake/absorption, chronic blood loss.	Serum ferritin, TIBC, and CBC.	Low serum ferritin and iron; elevated TIBC, microcytic, and hypochromic anemia.	Iron supplementation and treating the underlying cause.
Thalassemia	Genetic mutations leading to reduced/absent globin chain synthesis.	Hemoglobin electrophoresis, CBC, and genetic testing.	Microcytic anemia, abnormal hemoglobin pattern, and genetic confirmation.	Regular blood transfusions, iron chelation, and bone marrow transplantation in severe cases.

Relevant laboratory tests for detection, diagnosis, management, and monitoring

Table 7. Laboratory tests related to iron metabolism, hemoglobin disorders, and porphyrias, and analytical methods.

Test	Analytical Methods Available	Applied Aspects
1. Delta-aminolevulinic acid (ALA)	- Spectrophotometry - HPLC	- Diagnosis of acute porphyrias. - Monitoring of porphyria treatment.
2. Ferritin	- Immunoassay - Chemiluminescence	- Assessment of iron stores. - Diagnosis of iron deficiency and iron overload conditions.
3. Complete blood count (CBC)	- Automated hematology analyzers	- General health assessment. - Diagnosis of anemia, infection, and many other conditions.
4. G6PD	- Enzyme activity assay - Genetic testing	- Diagnosis of G6PD deficiency - Risk assessment for hemolytic anemia.
5. Haptoglobin	- Immunoturbidimetry - Nephelometry	- Detection of hemolytic anemia - Monitoring of hemolysis.
6. Hemoglobin	- Spectrophotometry - Hemoglobin electrophoresis	- Measurement of hemoglobin levels - Diagnosis of anemia.
7. Hemoglobin variants and thalassemias	- Hemoglobin electrophoresis – HPLC - Genetic testing	- Diagnosis and classification of hemoglobinopathies and thalassemias. - Genetic counseling.
8. Hemopexin	- Immunoassay - ELISA	- Detection of hemolytic anemia. - Differentiation from other causes of hemolysis.
9. Iron	- Colorimetric assay - Atomic absorption spectrometry	- Diagnosis of iron deficiency and iron overload - Monitoring of iron therapy.
10. Porphobilinogen (PBG)	- Watson-Schwartz test - Ion-exchange chromatography	- Diagnosis of acute porphyrias. - Monitoring of porphyria treatment.
11. Porphyrins	- Spectrophotometry - HPLC - Mass spectrometry	- Diagnosis of porphyrias - Monitoring of porphyria treatment.
12. Soluble transferrin receptor	- Immunoassay - ELISA	- Assessment of iron status - Differentiation between iron deficiency anemia and anemia of chronic disease.
13. Total iron binding capacity (TIBC)	- Colorimetric assay - Immunoassay	- Assessment of iron status - Diagnosis of iron deficiency and iron overload.
14. Transferrin	- Immunoassay - Nephelometry	- Assessment of iron status - Diagnosis of iron deficiency and iron overload.

(Table 7) cont.....

Test	Analytical Methods Available	Applied Aspects
15. Transferrin saturation	- Calculation from serum iron and TIBC	- Assessment of iron status - Diagnosis of iron deficiency and iron overload.

Biochemical basis, relevant lab tests, interpretation, and applied aspects

Table 8. Laboratory tests related to iron metabolism, hemoglobin disorders, and porphyrias, interpretation.

Test	Interpretation	Clinical Signs and Symptoms
1. Delta-aminolevulinic acid (ALA)	Elevated levels indicate acute porphyria.	Abdominal pain, neurological symptoms, and psychiatric symptoms.
2. Ferritin	Low levels indicate iron deficiency. High levels indicate iron overload or inflammation.	Fatigue, weakness (iron deficiency); liver disease, diabetes (iron overload).
3. Complete blood count (CBC)	Abnormal values indicate various conditions (*e.g.*, anemia, infection, leukemia).	Fatigue, pallor, bruising, infection, and bleeding.
4. G6PD	Low enzyme activity indicates G6PD deficiency.	Hemolytic anemia, jaundice, and dark urine after exposure to triggers.
5. Haptoglobin	Low levels indicate hemolysis. High levels may indicate inflammation.	Fatigue, pallor, and jaundice (hemolysis).
6. Hemoglobin	Low levels indicate anemia. High levels indicate Polycythemia.	Fatigue, weakness, and shortness of breath (anemia); headache and dizziness (polycythemia).
7. Hemoglobin variants and thalassemias	Abnormal hemoglobin patterns indicate specific hemoglobinopathies.	Fatigue, pallor, jaundice, splenomegaly, and skeletal abnormalities (thalassemia).
8. Hemopexin	Low levels indicate hemolysis.	Fatigue, pallor, and jaundice.
9. Iron	Low levels indicate iron deficiency. High levels indicate iron overload.	Fatigue and weakness (iron deficiency); liver disease and diabetes (iron overload).
10. Porphobilinogen (PBG)	Elevated levels indicate acute porphyria.	Abdominal pain, neurological symptoms, and psychiatric symptoms.
11. Porphyrins	Elevated levels indicate porphyria.	Photosensitivity, abdominal pain, and neurological symptoms.
12. Soluble transferrin receptor	Elevated levels indicate iron deficiency. Normal or low levels indicate anemia of chronic disease.	Fatigue and weakness.

(Table 8) cont.....

Test	Interpretation	Clinical Signs and Symptoms
13. Total iron binding capacity (TIBC)	Elevated levels indicate iron deficiency. Low levels indicate iron overload or inflammation.	Fatigue and weakness (iron deficiency); liver disease and diabetes (iron overload).
14. Transferrin	Low levels indicate malnutrition or liver disease. High levels indicate iron deficiency.	Fatigue and weakness (iron deficiency); jaundice and ascites (liver disease).
15. Transferrin saturation	Low levels indicate iron deficiency. High levels indicate iron overload.	Fatigue and weakness (iron deficiency); liver disease and diabetes (iron overload).

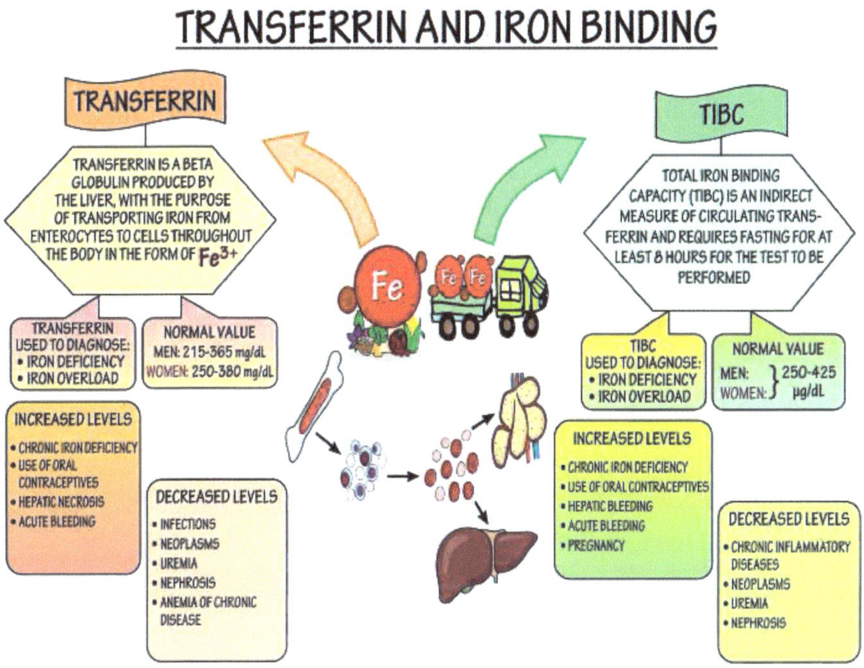

Fig. (1). Transferrin and iron binding.

SUMMARY

Iron is crucial for physiological processes, especially oxygen transport *via* hemoglobin. Disorders of iron metabolism, such as anemia and hemochromatosis, can profoundly affect health. Hemoglobinopathies, including sickle cell disease and thalassemia, result from genetic mutations affecting hemoglobin. Porphyrias, caused by defects in heme biosynthesis, lead to toxic intermediate accumulation and various clinical symptoms. This chapter provides an overview of these disorders and their management.

EXERCISES

1. Explain the role of hepcidin in iron metabolism and its implications in iron-related disorders.
2. Describe the genetic mutations involved in thalassemias and their impact on hemoglobin synthesis.
3. What are the common laboratory tests used to diagnose porphyrias, and how are the results interpreted?
4. Discuss the clinical management of iron overload in patients with hemochromatosis.
5. Analyze the pathophysiology of sickle cell disease and the laboratory tests used for its diagnosis.

CONCLUSION

The accurate diagnosis and effective management of iron and hemoglobin disorders, including porphyrias, require a deep understanding of their biochemical and genetic bases. Advances in laboratory diagnostics have improved the ability to detect and monitor these conditions, leading to better patient outcomes. Continued research and development of therapeutic strategies are essential for managing these complex disorders.

REFERENCES

[1] Warner, MJ.; Kamran, MT. Iron Deficiency Anemia. [Updated 2023 Aug 7]. In: StatPearls [Internet]. Treasure Island (FL): StatPearls Publishing. **2015**. Available from: https://www.ncbi.nlm.nih.gov/books/NBK448065/

[2] Bajwa, H, Basit H. Thalassemia. [Updated 2023 Aug 8]. In: StatPearls [Internet]. Treasure Island (FL): StatPearls Publishing. **2025**. Available from: https://www.ncbi.nlm.nih.gov/books/NBK545151/

[3] Ricci, A, Corradini E, Buzzetti E, Pietrangelo A, Ventura P. Modena Centre of Rare Diseases - Porphyria Working Group. Porphyrias: Pathophysiology and clinical management recommendations for hepatologists. *Hepatol Commun.,* **2025**, *9*(12), e0822. [http://dx.doi.org/10.1097/HC9.0000000000000822]

[4] Tahir, M. Sickle cell disease: understanding pathophysiology, clinical features and advances in gene therapy approaches. Front. Pharmacol. Sec. Experimental Pharmacology and Drug Discovery Volume 16 - **2025**. [http://dx.doi.org/10.3389/fphar.2025.1630994]

[5] Camaschella, C.; Nai, A.; Silvestri, L. Iron metabolism and iron disorders revisited in the hepcidin era. *Haematologica,* **2020**, *105*(2), 260-272. [http://dx.doi.org/10.3324/haematol.2019.232124]

CHAPTER 14

Lipids and Disorders of Lipoprotein Metabolism

Abstract: Disorders of lipoprotein metabolism encompass a range of conditions that significantly contribute to cardiovascular diseases. These disorders are often characterized by abnormalities in the levels and composition of plasma lipids and lipoproteins. The clinical relevance of these disorders is underscored by their association with atherosclerosis, a leading cause of morbidity and mortality worldwide. This chapter delves into the biochemical underpinnings of lipid disorders, exploring the diagnostic tools and therapeutic approaches employed in clinical practice. By understanding these mechanisms, clinicians and laboratory professionals can improve patient outcomes through targeted interventions and personalized medicine approaches.

Keywords: Lipid disorders, Lipoprotein metabolism, Plasma lipids, Plasma lipoproteins.

LEARNING OBJECTIVES

After going through the chapter, one should be able to:

- Describe the principles and control of lipid metabolism and key aspects of the study of lipid and lipoprotein disorders.
- Describe the causes, clinical signs, and symptoms of disorders relating to lipid and lipoprotein disorders.
- Describe biochemical and clinical laboratory methods for the detection, diagnosis, and management of disorders relating to lipid and lipoprotein disorders.
- Describe specific laboratory investigations important to the study of disorders of lipid and lipoprotein metabolism.
- Recognize analytical methods available for their measurement.
- Select laboratory investigations and interpret the analytical results in the context of clinical signs and symptoms.

Simmi Kharb

KEY CONCEPTS: LIPID METABOLISM AND ITS REGULATION

Principles and Control of Lipid Metabolism (Tables 1, 2)

Table 1. Lipid metabolism and lipoprotein disorders.

Key Concepts	Description	Applied Aspects
Overview of Lipid Metabolism	Involves the synthesis, storage, and breakdown of lipids, including fatty acids, triglycerides, phospholipids, and cholesterol, in the body	For managing metabolic diseases, such as obesity, diabetes, and cardiovascular diseases
Lipogenesis	Process of fatty acid synthesis from acetyl-CoA and malonyl-CoA by the enzyme fatty acid synthase Occurs in the liver and adipose tissue	Serve as a target for dietary interventions, anti-obesity drugs
Lipolysis	Breakdown of triglycerides into free fatty acids and glycerol by hormone-sensitive lipase Occurs in adipose tissue	Enhancing lipolysis can be beneficial in weight loss strategies and in treating metabolic syndrome
Beta-Oxidation	Breakdown of fatty acids in mitochondria to generate acetyl-CoA and ATP	For energy production, particularly during fasting or prolonged exercise Defects can lead to metabolic disorders
Ketogenesis	Production of Ketone bodies (acetoacetate, beta-hydroxybutyrate, and acetone) from acetyl-CoA in liver during periods of low carbohydrate intake	Ketogenic diets utilize this pathway for weight loss and managing epilepsy Excessive ketogenesis can lead to ketoacidosis
Cholesterol Metabolism	Involves the synthesis, absorption, and elimination of cholesterol. Key pathways: mevalonate pathway and conversion of cholesterol to bile acids	Drug targets: statins and other cholesterol-lowering drugs to reduce cardiovascular disease risk
Lipoprotein Metabolism	Lipoproteins transport lipids in the blood, and each type has a specific role in lipid transport and metabolism: Includes chylomicrons, VLDL, LDL, and HDL (Fig. **1**)	Alterations in lipoprotein metabolism are implicated in atherosclerosis Targets to modify lipoprotein levels and improve cardiovascular outcomes
Regulation of Lipid Metabolism	Regulation: Hormonal: insulin, glucagon, and epinephrine Dietary factors Genetic regulation (*e.g.*, SREBP, PPARs)	Target for interventions: changes (diet, exercise) Pharmacological agents (*e.g.*, fibrates, thiazolidinediones) Gene therapy approaches

(Table 1) cont.....

Key Concepts	Description	Applied Aspects
Fatty Acid Synthesis Pathway	Key enzymes: Acetyl-CoA Carboxylase (ACC) and Fatty Acid Synthase (FAS) Upregulated in the fed state Downregulated in the fasting state	Target for anti-obesity and anti-cancer drugs
Fatty Acid Oxidation Pathway	Key regulatory enzyme: Palmitoyl transferase I (CPT1): Controls the entry of fatty acids into mitochondria for oxidation	Deficiencies can lead to fatty acid oxidation disorders Therapeutic targets: dietary management and enzyme replacement therapies
Triglyceride Storage and Mobilization	Storage form in adipose tissue Mobilization regulated by hormonal signals: insulin and catecholamines	Important in energy balance and obesity Therapeutic target: interventions to manage excessive triglyceride storage or enhance mobilization in metabolic diseases
Impact of Diet and Exercise	Dietary intake of fats influences lipid metabolism Exercise increases fatty acid oxidation and improves lipid profile	Target for strategies in managing dyslipidemia and preventing cardiovascular diseases: Nutritional interventions and physical activity
Genetic Factors in Lipid Metabolism	Genetic variations (*e.g.*, ApoE polymorphisms) affect lipid levels, posing the risk of cardiovascular diseases	Personalized treatment strategies for dyslipidemia and prevention of disease
Pharmacological Regulation	Drug target: Statins, fibrates, niacin, and omega-3 fatty acids are used to manage dyslipidemia Acts by various mechanisms to reduce LDL and VLDL and increase HDL	Management of hyperlipidemia Prevention of atherosclerotic cardiovascular disease
Metabolic Syndrome and Lipid Metabolism	Cluster of conditions: Insulin resistance, hypertension, and dyslipidemia Often associated with obesity	Target for comprehensive management: Includes lifestyle interventions, pharmacotherapy, and monitoring to reduce cardiovascular risk
Non-Alcoholic Fatty Liver Disease (NAFLD)	Excessive accumulation of fat in the liver, not due to alcohol consumption Often associated with obesity, diabetes, and metabolic syndrome	Target management includes weight loss, insulin sensitizers, and lipid-lowering agents Prevention focuses on lifestyle modifications

Table 2. Lipid and lipoprotein disorders: apolipoproteins.

Apolipoprotein	Functions	Receptors	Applied Aspects
Apolipoprotein A-I (ApoA-I)	- Major protein component of HDL - Activates Lecithin-Cholesterol Acyltransferase (LCAT), which is crucial for HDL maturation - Promotes cholesterol efflux from tissues to the liver for excretion	- HDL receptor (SR-BI)	- Critical for reverse cholesterol transport - High levels are associated with reduced risk of cardiovascular diseases - Target for therapies aimed at increasing HDL levels
Apolipoprotein B-100 (ApoB-100)	- Main protein component of LDL, VLDL, and IDL - Essential for the assembly and secretion of VLDL from the liver - Mediates LDL binding to the LDL-Receptor (LDL-R)	- LDL receptor (LDL-R)	- Integral in the development of atherosclerosis - Elevated levels are a risk factor for cardiovascular diseases - Target for lipid-lowering therapies like statins
Apolipoprotein B-48 (ApoB-48)	- Structural protein of chylomicrons - Essential for the assembly and secretion of chylomicrons from the intestine	- N/A	- Important for the transport of dietary lipids from the intestines to other tissues
Apolipoprotein C-II (ApoC-II)	- Activator of Lipoprotein Lipase (LPL), which hydrolyzes triglycerides in VLDL and chylomicrons	- N/A	- Deficiency leads to hypertriglyceridemia - Potential therapeutic target for managing hypertriglyceridemia
Apolipoprotein C-III (ApoC-III)	- Inhibits Lipoprotein Lipase (LPL) and hepatic uptake of triglyceride-rich particles	- N/A	- High levels are associated with hypertriglyceridemia and increased risk of cardiovascular diseases - Potential target for reducing triglyceride levels
Apolipoprotein E (ApoE)	- Mediates the clearance of chylomicron remnants and VLDL from the bloodstream by binding to LDL-Receptor (LDL-R) and LDL Receptor-related Protein (LRP).	- LDL receptor (LDL-R) - LDL receptor-related protein (LRP)	- Genetic variants (*e.g.*, ApoE4) are associated with increased risk of Alzheimer's disease and cardiovascular diseases - Target for therapies aimed at modifying lipid metabolism and reducing disease risk
Apolipoprotein A-II (ApoA-II)	- Structural protein of HDL - Modulates HDL metabolism and function	- HDL receptor (SR-BI)	- Role in HDL function and reverse cholesterol transport - Potential target for therapies aimed at modifying HDL metabolism

(Table 2) cont.....

Apolipoprotein	Functions	Receptors	Applied Aspects
Apolipoprotein A-IV (ApoA-IV)	- Involved in the regulation of food intake and lipid metabolism - Acts as a cofactor for LCAT	- N/A	- Potential role in satiety and obesity management - May have anti-atherogenic properties.
Apolipoprotein D (ApoD)	- Associated with HDL and LDL - Involved in lipid transport and metabolism	- N/A	- Potential role in neuroprotection and lipid metabolism.
Apolipoprotein H (ApoH, β2-Glycoprotein I)	- Involved in the regulation of coagulation - Binds to negatively charged phospholipids and proteins on the surface of damaged cells	- N/A	- Implicated in antiphospholipid syndrome - Potential biomarker and therapeutic target in autoimmune and coagulation disorders

KEY CONCEPTS: LIPID AND LIPOPROTEIN DISORDERS

Cardiovascular Disease Risk Calculation and Evaluation

Disorders Related to Lipid and Lipoprotein (Tables 3 - 6) [1 - 5]

Table 3. Disorders related to lipid and lipoprotein: cardiovascular disorders.

Key Concepts	Description	Applied Aspects
Risk Factors	Include age, gender, smoking, hypertension, diabetes, total cholesterol, LDL, HDL, and family history	- Identification of high-risk individuals for preventive measures - Lifestyle modifications and pharmacotherapy to manage risk factors
Risk Scores	Tools like Framingham Risk Score, ASCVD Risk Calculator, and QRISK, used to estimate 10-year CVD risk	- Guides clinical decisions on statin use, aspirin therapy, and lifestyle changes
Cost-Effectiveness of Lipid Screening	Assessment of the value of routine lipid screening in different populations	- Screening strategies tailored to high-risk populations can be more cost-effective - Policy implications for health insurance coverage and preventive care programs

KEY CONCEPTS: FATTY ACID TRANSPORT AND OXIDATION

Table 4. Disorders related to fatty acid transport and oxidation.

Key Concepts	Description	Applied Aspects
Transport	Fatty acids are transported in the blood bound to albumin or as part of lipoproteins	- Targeting transport mechanisms in obesity and metabolic syndrome
Oxidation	Beta-oxidation in mitochondria breaks down fatty acids to produce acetyl-CoA, NADH, and FADH2	- Critical for energy production during fasting and exercise - Defects in oxidation can lead to metabolic disorders; therapeutic targets include enzyme replacement and dietary management

KEY CONCEPTS: LIPID ABSORPTION, TRANSPORT, AND METABOLISM

Table 5. Disorders related to lipid absorption, transport, and metabolism.

Key Concepts	Description	Applied Aspects
Absorption	Dietary lipids are emulsified by bile, digested by pancreatic lipase, and absorbed as micelles in the intestines	- Disorders in lipid absorption can lead to malnutrition and fat-soluble vitamin deficiencies - Therapeutic interventions include enzyme supplements and dietary modifications
Transport	Lipids are transported as chylomicrons, VLDL, LDL, and HDL in the bloodstream	- Key to understanding lipid-related diseases like atherosclerosis - Targeting transport pathways can help manage dyslipidemia and reduce cardiovascular risk
Metabolism	Involves the synthesis, storage, and breakdown of lipids in tissues	- Metabolic disorders such as obesity, diabetes, and fatty liver disease - Therapeutic approaches include lifestyle changes, pharmacotherapy, and bariatric surgery

Table 6. Disorders related to lipoprotein metabolism.

Pathway	Description	Applied Aspects
Exogenous Pathway	Involves the transport of dietary lipids from the intestines to other tissues *via* chylomicrons	- Understanding fat absorption and transport can help manage conditions like hyperlipidemia and fat malabsorption syndromes
Endogenous Pathway	Involves the transport of lipids synthesized in the liver to other tissues *via* VLDL, IDL, and LDL	- Crucial for managing blood lipid levels and preventing atherosclerosis - Statins and other drugs target this pathway to reduce LDL levels

KEY CONCEPTS: LIPOPROTEIN METABOLISM: ENDOGENOUS AND EXOGENOUS PATHWAYS

Biochemical Basis, Relevant Lab Tests, Interpretation, and Applied Aspects

Biochemical and Clinical Laboratory Methods for Detection, Diagnosis, and Management of Disorders Relating to Lipid and Lipoprotein Disorders (Tables 7-11) [1 - 5]

Table 7. Atherosclerosis.

Aspect	Details
Biochemical Basis	- Accumulation of lipids, cholesterol, and inflammatory cells in arterial walls - Formation of atherosclerotic plaques leading to narrowed and hardened arteries
Relevant Lab Tests	- Lipid panel (total cholesterol, LDL, HDL, triglycerides) - High-sensitivity C-Reactive Protein (hs-CRP) - Apolipoproteins (ApoA-I, ApoB)
Interpretation	- Elevated LDL, total cholesterol, and hs-CRP indicate a higher risk - Low HDL: risk factor
Applied Aspects	- Lifestyle changes: diet, exercise - Drug targets: statins, PCSK9 inhibitors - Monitoring and managing blood pressure and blood glucose

Table 8. Hypercholesterolemia.

Aspect	Details
Biochemical Basis	- Elevated levels of cholesterol in the blood, primarily due to high levels of LDL
Relevant Lab Tests	- Lipid panel: total cholesterol, LDL, HDL, triglycerides - Genetic testing for familial hypercholesterolemia
Interpretation	- LDL > 160 mg/dL - Genetic mutations (*e.g.*, LDLR, APOB, PCSK9) confirm inherited forms
Applied Aspects	- Lifestyle modifications (diet, exercise) - Statins, ezetimibe, and PCSK9 inhibitors for pharmacotherapy - Regular monitoring of lipid levels

Table 9. Hyperlipidemia.

Aspect	Inherited Disorders	Non-Inherited Disorders
Biochemical Basis	- Genetic mutations affecting lipid metabolism *e.g.*, familial hypercholesterolemia, familial combined hyperlipidemia	- Secondary causes such as diet, obesity, diabetes, hypothyroidism, nephrotic syndrome, and certain medications

(Table 9) cont.....

Aspect	Inherited Disorders	Non-Inherited Disorders
Relevant Lab Tests	- Lipid panel (total cholesterol, LDL, HDL, triglycerides) - Genetic testing for specific mutations (*e.g.*, LDLR, APOB, PCSK9)	- Lipid panel: total cholesterol, LDL, HDL, triglycerides - Tests to identify secondary causes: thyroid function tests, glucose levels
Interpretation	- High LDL or triglycerides with genetic mutations confirm inherited forms - HDL levels may also be affected	- High LDL or triglycerides without genetic mutations indicate non-inherited forms - Assess for secondary causes based on additional test results
Applied Aspects	- Early detection and family screening - Lifestyle changes and pharmacotherapy tailored to specific genetic forms - Regular monitoring of lipid levels	- Lifestyle changes and treatment of underlying conditions - Pharmacotherapy: Statins, fibrates, niacin, and omega-3 fatty acids

Table 10. Hypolipidemia.

Aspect	Details
Biochemical Basis	- Abnormally low levels of lipids in the blood, potentially due to genetic factors (*e.g.*, abetalipoproteinemia) or secondary causes (*e.g.*, malnutrition, hyperthyroidism)
Relevant Lab Tests	- Lipid panel (total cholesterol, LDL, HDL, triglycerides) - Genetic testing for specific disorders (*e.g.*, abetalipoproteinemia, hypobetalipoproteinemia)
Interpretation	- Total cholesterol < 120 mg/dL, LDL < 50 mg/dL, triglycerides < 50 mg/dL - Genetic mutations confirm inherited forms
Applied Aspects	- Identify and treat underlying causes - Nutritional support and management of associated conditions - Regular monitoring of lipid levels and overall health

Table 11. Metabolic syndrome.

Aspect	Details
Biochemical Basis	- Cluster of conditions including insulin resistance, hypertension, dyslipidemia (high triglycerides, low HDL), and central obesity
Relevant Lab Tests	- Lipid panel: total cholesterol, LDL, HDL, and triglycerides - Fasting blood glucose or HbA1c - Blood pressure measurements - Waist circumference
Interpretation	- Presence of three or more criteria: elevated waist circumference, high triglycerides, low HDL, high blood pressure, high fasting glucose

(Table 11) cont.....

Aspect	Details
Applied Aspects	- Lifestyle modifications: diet, exercise - Pharmacotherapy: statins, antihypertensives, antidiabetics - Regular monitoring and management of individual risk factors.

Relevant Laboratory Tests, Analytics Methods Available, and Applied Aspects

Specific Laboratory Investigations to Study the Disorders of Lipid and Lipoprotein Metabolism: Interpretaion, Analytical Methods available (Tables 12-22; Fig. 1) [1 - 5]

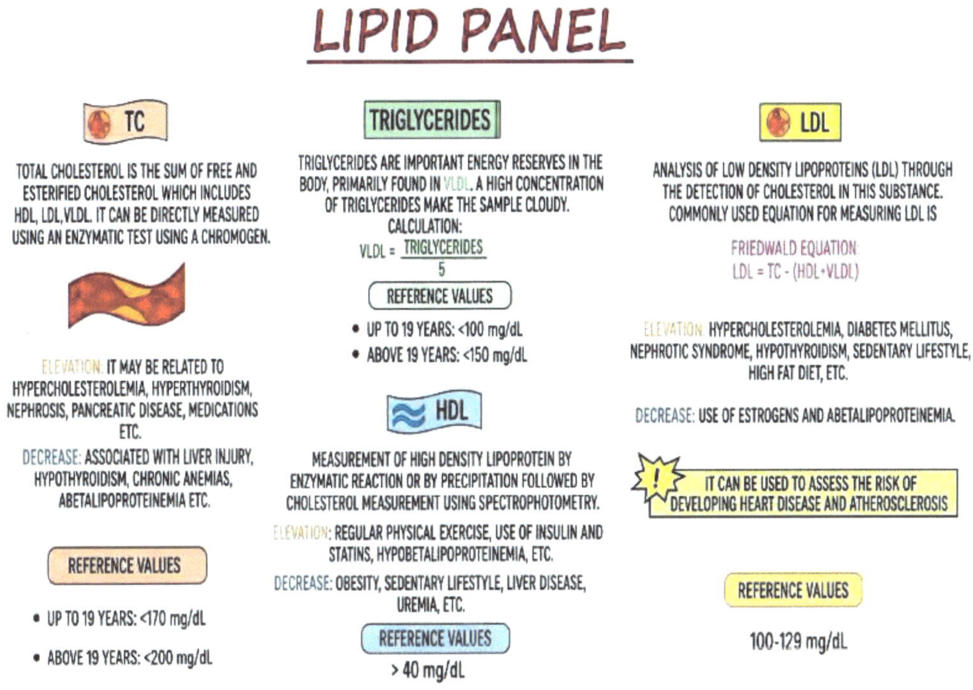

Fig. (1). Lipid panels.

Table 12. Total cholesterol.

Aspect	Details
Test	Total Cholesterol
Analytical Methods	- Enzymatic colorimetric assay

(Table 12) cont.....

Aspect	Details
Interpretation	- Desirable: < 200 mg/dL - Borderline high: 200-239 mg/dL - High: ≥ 240 mg/dL
Applied Aspects	- Used to assess cardiovascular risk - Guides therapeutic decisions on lifestyle changes and lipid-lowering medications.

Table 13. Triglycerides.

Aspect	Details
Test	Triglycerides
Analytical Methods	- Enzymatic colorimetric assay
Interpretation	- Normal: < 150 mg/dL - Borderline high: 150-199 mg/dL - High: 200-499 mg/dL - Very high: ≥ 500 mg/dL
Applied Aspects	- High levels associated with increased risk of pancreatitis and cardiovascular disease - Management includes lifestyle changes and triglyceride-lowering medications

Table 14. Apolipoprotein A and B.

Aspect	Details
Test	Apolipoprotein A (ApoA-I) and Apolipoprotein B (ApoB)
Analytical Methods	- Immunoassay (*e.g.*, nephelometry, turbidimetry)
Interpretation	- ApoA-I: Reflect HDL levels; higher levels are protective - ApoB: Reflects LDL levels; higher levels indicate increased cardiovascular risk
Applied Aspects	- Provides more precise cardiovascular risk assessment compared to standard lipid panels - Guides therapeutic decisions and monitoring of lipid-lowering therapy.

Table 15. Genotyping of ApoE, LDL Receptor, and LPL.

Aspect	Details
Test	Genotyping for ApoE, LDL-Receptor (LDLR), and Lipoprotein Lipase (LPL) genes
Analytical Methods	- PCR followed by sequencing or Restriction Fragment Length Polymorphism (RFLP) analysis
Interpretation	- ApoE: E4 variant associated with increased cardiovascular and Alzheimer's risk - LDLR: Mutations cause familial hypercholesterolemia - LPL: Mutations affect triglyceride metabolism

Aspect	Details
Applied Aspects	- Identifies individuals with a genetic predisposition to dyslipidemias - Personalizes treatment strategies based on genetic profile.

Table 16. HDL cholesterol.

Aspect	Details
Test	HDL Cholesterol
Analytical Methods	- Direct enzymatic assay - Precipitation method
Interpretation	- Low: < 40 mg/dL (men), < 50 mg/dL (women) - High: ≥ 60 mg/dL
Applied Aspects	- High levels are protective against cardiovascular disease -Management: lifestyle changes to increase HDL levels.

Table 17. High-sensitivity CRP (hs-CRP).

Aspect	Details
Test	High-Sensitivity C-Reactive Protein (hs-CRP)
Analytical Methods	- Immunoassay (*e.g.*, ELISA, nephelometry)
Interpretation	- Low risk: < 1.0 mg/L - Average risk: 1.0-3.0 mg/L - High risk: > 3.0 mg/L
Applied Aspects	- Marker of inflammation associated with cardiovascular risk - Used to refine cardiovascular risk assessment and guide preventive strategies.

Table 18. LDL cholesterol – direct and calculated methods (friedewald equation).

Aspect	Details
Test	LDL Cholesterol
Analytical Methods	- Direct enzymatic assay - Calculated using Friedewald equation: LDL = Total cholesterol - HDL - (Triglycerides/5)
Interpretation	- Optimal: < 100 mg/dL - Near optimal/above optimal: 100-129 mg/dL - Borderline high: 130-159 mg/dL - High: 160-189 mg/dL - Very high: ≥ 190 mg/dL
Applied Aspects	- Direct method used when triglycerides > 400 mg/dL - LDL: primary target for lipid-lowering therapy

Table 19. Lipoprotein electrophoresis.

Aspect	Details
Test	Lipoprotein Electrophoresis
Analytical Methods	- Electrophoretic separation of lipoproteins on agarose or polyacrylamide gel
Interpretation	- Identifies and quantifies lipoprotein fractions (chylomicrons, VLDL, LDL, HDL)
Applied Aspects	- Used to diagnose and characterize lipid disorders - Helps in understanding the pattern of dyslipidemia and tailoring treatment

Table 20. Lipoprotein ultracentrifugation.

Aspect	Details
Test	Lipoprotein Ultracentrifugation
Analytical Methods	- Density gradient ultracentrifugation
Interpretation	- Separates lipoproteins based on density (chylomicrons, VLDL, IDL, LDL, HDL)
Applied Aspects	- Gold standard for detailed lipoprotein profiling - Provides precise data for research and complex clinical cases

Table 21. Lp (a).

Aspect	Details
Test	Lipoprotein(a) [Lp (a)]
Analytical Methods	- Immunoassay (*e.g.*, ELISA, nephelometry)
Interpretation	- Elevated Lp (a) levels (>30 mg/dL) associated with increased cardiovascular risk
Applied Aspects	- Genetic risk factor for atherosclerosis - Monitoring and managing elevated Lp (a) levels in high-risk patients.

Table 22. Non-HDL cholesterol (calculated).

Aspect	Details
Test	Non-HDL Cholesterol
Analytical Methods	- Calculated: Total cholesterol - HDL cholesterol
Interpretation	- Non-HDL cholesterol includes all atherogenic lipoproteins (LDL, VLDL, IDL).
Applied Aspects	- Better predictor of cardiovascular risk than LDL alone - Target for therapy in patients with high triglycerides or metabolic syndrome

SUMMARY

Lipoprotein metabolism disorders, such as hypercholesterolemia, hypertriglyceridemia, and mixed dyslipidemias, play a crucial role in the development of atherosclerotic cardiovascular disease. These disorders result from genetic and lifestyle factors that disrupt normal lipid and lipoprotein metabolism. This chapter offers a comprehensive analysis of the biochemical pathways involved in lipid metabolism and outlines the diagnostic criteria for different lipid disorders, including advanced lipid testing and genetic screening.

EXERCISES

- Discuss briefly the enzyme playing a critical role in the hydrolysis of triglycerides in VLDL and chylomicrons.
- Which lipoprotein is most associated with atherogenesis and cardiovascular risk, and how?
- Patients with hypertriglyceridemia are at risk for which complications?
- Describe the clinical significance of LDL-C in the assessment of cardiovascular risk.
- Discuss the mechanism by which statins reduce plasma LDL levels.
- Discuss the diagnosis and lab investigations for a 50-year-old patient presenting with high LDL and HDL levels, having a family history of cardiovascular disease.

CONCLUSION

Studying lipids and disorders of lipoprotein metabolism is crucial for preventing and managing cardiovascular diseases. Advances in diagnostic techniques and therapeutic agents have improved the outlook for patients. Further research is necessary to understand the genetic and molecular mechanisms behind these conditions for more effective treatments. Laboratory professionals play a critical role in diagnosis and monitoring therapeutic outcomes.

ACKNOWLEDGEMENT AND CONFLICT OF INTEREST

No potential conflict of interest is declared by the author. It is also declared that the complete work is an individual effort by the author and that no financial/ administrative/ academic support was availed from any individual/ institution, or organization.

REFERENCES

[1] Third Report of the National Cholesterol Education Program (NCEP) Expert Panel on Detection, Evaluation, and Treatment of High Blood Cholesterol in Adults (Adult Treatment Panel III) Final Report. *Circulation,* **2002**, *106*(25), 3143-3421.

[http://dx.doi.org/10.1161/circ.106.25.3143] [PMID: 12485966]

[2] Grundy, S.M.; Stone, N.J.; Bailey, A.L.; Beam, C.; Birtcher, K.K.; Blumenthal, R.S.; Braun, L.T.; de Ferranti, S.; Faiella-Tommasino, J.; Forman, D.E.; Goldberg, R.; Heidenreich, P.A.; Hlatky, M.A.; Jones, D.W.; Lloyd-Jones, D.; Lopez-Pajares, N.; Ndumele, C.E.; Orringer, C.E.; Peralta, C.A.; Saseen, J.J.; Smith, S.C., Jr; Sperling, L.; Virani, S.S.; Yeboah, J. 2018 AHA/ACC/AACVPR/AAPA/ABC/ACPM/ADA/AGS/APhA/ASPC/NLA/PCNA Guideline on the Management of Blood Cholesterol. *J. Am. Coll. Cardiol.,* **2019**, *73*(24), e285-e350. [http://dx.doi.org/10.1016/j.jacc.2018.11.003] [PMID: 30423393]

[3] Alam, U.; Mathai, S.V.; Filtz, A.; Kuno, T.; Badimon, J.J.; Sniderman, A.D.; Virani, S.S.; Toth, P.P.; Shapiro, M.D.; Lavie, C.J.; Bhatt, D.L.; Slipczuk, L. Targeting triglycerides in cardiovascular disease prevention: evidence, mechanisms, and emerging therapies. *Curr Cardiol Rep,* **2026**, *28*(1), 8. [http://dx.doi.org/10.1007/s11886-025-02337-1]

[4] Duan, Y.; Gong, K.; Xu, S. Regulation of cholesterol homeostasis in health and diseases: from mechanisms to targeted therapeutics. *Sig Transduct Target Ther,* **2022**, *7*, 265. [http://dx.doi.org/10.1038/s41392-022-01125-5]

[5] Shapiro, MD. Between the extremes: the impact of lipid variability on cardiovascular risk. Circ Cardiovasc Imaging 2024; Volume 17, Number 9. [http://dx.doi.org/10.1161/CIRCIMAGING.124.01737]

CHAPTER 15

Minerals: Calcium, Magnesium, Parathyroid, Bone Disorders

Abstract: Calcium and magnesium play essential roles in neuromuscular function, bone health, and metabolic processes, and are regulated by hormones such as Parathyroid Hormone (PTH) and vitamin D. Dysregulation of these minerals can lead to disorders such as hypocalcemia, hypercalcemia, hypomagnesemia, and bone-related conditions like osteoporosis and Paget's disease. This paper explores the applied clinical chemistry aspects of mineral metabolism, focusing on diagnostic methods, interpretation of lab results, and treatment options for calcium, magnesium, and bone disorders.

Keywords: Bone disorders, Calcium, Magnesium, Parathyroid hormone, Vitamin D.

LEARNING OBJECTIVES

After reading this chapter, one should be able to:

- Describe the control of calcium and phosphate homeostasis.
- Describe the causes, clinical signs, and symptoms of the disorders and manifestations of calcium and phosphate metabolism and bone disease.
- Describe how laboratory investigations are important in their detection, diagnosis, and management.
- Describe the specific laboratory investigations important to studying calcium and phosphate metabolism and bone disease.
- Recognize the analytical methods available for their measurement.
- Select laboratory investigations and interpret the analytical results in the context of clinical signs and symptoms.

Simmi Kharb

KEY CONCEPTS: CALCIUM AND PHOSPHATE HOMEOSTASIS

Control of Calcium and Phosphate Homeostasis (Table 1)

Table 1. Control of calcium and phosphate homeostasis.

Mineral	Key Concepts	Applied Aspects
Calcium	Circulating forms of calcium: - Ionized Calcium (50%): Active form, free in plasma. - Protein-bound Calcium (40%): Mostly bound to albumin. - Complexed Calcium (10%): Bound to anions like phosphate.	- Ionized calcium: Physiologically active form Crucial for: muscle contraction nerve conduction clotting - Protein-bound calcium: levels are affected by albumin concentration, making correction important in hypoalbuminemia.
Vitamin D	- Vitamin D3 (cholecalciferol): Synthesized in the skin from sunlight. - Hydroxylation in the liver produces 25-hydroxyvitamin D. - Kidney hydroxylation forms active 1,25-dihydroxyvitamin D (calcitriol).	- Calcitriol: Increases calcium and phosphate absorption in the intestines and regulates bone resorption. - Deficiency causes rickets in children and osteomalacia in adults. - Lab assays: Measure 25(OH)D to assess vitamin D status.
Markers of Bone Resorption and Formation	- Bone Resorption Markers: CTX (C-terminal telopeptide), NTX (N-terminal telopeptide), and TRAP (tartrate-resistant acid phosphatase). - Bone Formation Markers: ALP (alkaline phosphatase), P1NP (procollagen type I N-terminal propeptide), and osteocalcin.	- Elevated bone resorption markers indicate conditions: Osteoporosis and Paget's disease. - Bone formation markers: Elevated in growing children, healing fractures, and bone metabolic diseases. - Used to monitor response to therapies.
Disorders of calcium homeostasis	- Primary Hypercalcemia: Caused by conditions like hyperparathyroidism (increased PTH). - Secondary Hypercalcemia: Due to extrinsic factors: excessive vitamin D or malignancies. - Primary Hypocalcemia: Result of insufficient PTH (*e.g.*, hypoparathyroidism). - Secondary Hypocalcemia: Caused by: Chronic kidney disease, leading to impaired vitamin D metabolism.	- Diagnosis: Measurement of PTH, calcium, phosphate, and vitamin D levels. - Primary Hypercalcemia: Managed by parathyroidectomy or medications. - Secondary Hypercalcemia: Treated by addressing the underlying cause (*e.g.*, cancer, vitamin D toxicity).

(Table 1) cont.....

Mineral	Key Concepts	Applied Aspects
Calcium and phosphate homeostasis	- Regulated by: Parathyroid Hormone (PTH), vitamin D, and calcitonin. - PTH: Increases calcium reabsorption in the kidneys, stimulates bone resorption, and activates vitamin D. - Vitamin D (calcitriol): Increases intestinal absorption of calcium and phosphate. - Calcitonin: Lowers blood calcium by inhibiting bone resorption.	- PTH and vitamin D work in concert to maintain calcium homeostasis. - Disorders, namely hypoparathyroidism, vitamin D deficiency, or chronic kidney disease, disrupt this regulation. - Correcting imbalances may involve supplementation or hormone therapy.

Biochemical Basis, Relevant Lab Tests, Interpretation, and Applied Aspects

Laboratory Investigations in Detection, Diagnosis, and Management of Disorders (Table 2)

Table 2. Biochemical basis, relevant lab tests, interpretation, and applied aspects.

Condition	Biochemical Basis	Relevant Lab Tests	Interpretation of Results	Applied Aspects
Hypercalcemia	Excess calcium in the blood due to increased bone resorption, excessive vitamin D, or PTH production.	Serum calcium (total/ionized), PTH, vitamin D, phosphate, and ALP	Elevated calcium, high or low PTH, low phosphate, and high ALP in bone-related causes	Requires differentiation between primary hyperparathyroidism, malignancy, and vitamin D toxicity. Treatment depends on the underlying cause.
Hypermagnesemia	Excess magnesium due to renal failure, excessive intake of magnesium supplements, or drugs.	Serum magnesium, calcium, PTH, and kidney function tests	Elevated magnesium, potential hypocalcemia	Often seen in renal failure or patients receiving magnesium-based treatments. Treated by restricting intake or dialysis.
Hyperpara-thyroidism	Overproduction of PTH, leading to increased calcium and decreased phosphate levels.	Serum calcium, PTH, phosphate, ALP, and vitamin D	Elevated calcium, high PTH, and low phosphate	Primary hyperparathyroidism, treated with surgery; secondary hyperparathyroidism, managed by treating the underlying cause (*e.g.*, CKD).

(Table 2) cont.....

Condition	Biochemical Basis	Relevant Lab Tests	Interpretation of Results	Applied Aspects
Hyperphosphatemia	Elevated phosphate levels, often due to kidney dysfunction, increased intake, or cellular breakdown (*e.g.*, rhabdomyolysis).	Serum phosphate, calcium, PTH, renal function tests, and vitamin D	High phosphate, potential hypocalcemia, and elevated PTH in secondary hyperparathyroidism	Common in CKD. Managed by phosphate binders and dietary restriction.
Hypocalcemia	Low calcium levels caused by vitamin D deficiency, hypoparathyroidism, or renal failure.	Serum calcium (total/ionized), PTH, phosphate, and vitamin D	Low calcium, low or high PTH (depending on cause), and high phosphate in hypoparathyroidism	Treatment includes calcium and vitamin D supplementation, addressing the underlying cause.
Hypomagnesemia	Low magnesium levels due to poor intake, GI losses, or renal losses.	Serum magnesium, calcium, and potassium	Low magnesium, possible hypocalcemia, and hypokalemia	Magnesium supplements or correcting the underlying cause (*e.g.*, GI/renal losses).
Hypoparathyroidism	Deficient PTH production leads to low calcium and high phosphate.	Serum calcium, PTH, phosphate, magnesium, and vitamin D	Low calcium, low PTH, and high phosphate	Managed with calcium and vitamin D supplementation, sometimes PTH analogs.
Hypophosphatemia	Low phosphate levels caused by malnutrition, refeeding syndrome, or chronic alcohol use.	Serum phosphate, calcium, PTH, ALP, and vitamin D	Low phosphate and potential elevated PTH	Corrected by oral or IV phosphate supplementation, treating the underlying cause (*e.g.*, malnutrition).
Hypophosphatasia	Genetic disorder characterized by deficient alkaline phosphatase activity, leading to abnormal bone mineralization.	Serum ALP, calcium, phosphate, PTH, and vitamin D	Low ALP, normal to elevated calcium and phosphate	Treatment may include enzyme replacement therapy.
Pseudohypoparathyroidism	Resistance to PTH, leading to hypocalcemia and hyperphosphatemia despite high PTH levels.	Serum calcium, PTH, phosphate, and ALP	Low calcium, high PTH, and high phosphate	Managed with calcium and vitamin D supplementation, often lifelong.
Osteogenesis Imperfecta	A genetic disorder causing defective collagen synthesis, leading to fragile bones.	Serum calcium, phosphate, PTH, ALP, and collagen biomarkers	Normal calcium and phosphate, increased ALP in the bone turnover stages	Treatment includes bisphosphonates, physiotherapy, and surgical interventions for fractures.

(Table 2) cont.....

Condition	Biochemical Basis	Relevant Lab Tests	Interpretation of Results	Applied Aspects
Osteoporosis	Decreased bone mass and strength, often due to hormonal changes (*e.g.*, menopause) or aging.	DEXA scan, serum calcium, phosphate, ALP, PTH, and vitamin D	Normal serum calcium and phosphate, and elevated ALP during bone turnover	Treated with calcium, vitamin D, bisphosphonates, or other osteoporosis medications.
Paget's Disease	Increased bone resorption and abnormal bone formation, leading to structurally abnormal bones.	Serum calcium, phosphate, ALP, and bone-specific ALP	Elevated ALP, normal to elevated calcium and phosphate	Managed with bisphosphonates and pain management approaches.
Hypercalcemia of Malignancy	Malignant tumors secrete PTH-related Protein (PTHrP) or cause direct bone destruction, leading to high calcium levels.	Serum calcium, PTH, PTHrP, vitamin D, and ALP	Elevated calcium, low PTH, and elevated PTHrP	Requires treatment of underlying malignancy, calcium-lowering drugs (*e.g.*, bisphosphonates, calcitonin), and hydration.

Relevant Laboratory Tests, Analytics Methods Available, and Applied Aspects

Laboratory Investigations for Studying Calcium and Phosphate Metabolism and Bone Disease and Available Analytical Methods (Table 3) [1 - 5]

Table 3. Relevant laboratory tests for calcium metabolism, analytics methods available, and applied aspects.

Test	Analytical Methods	Applied Aspects
1,25 di-hydroxy vitamin D	- Immunoassay: ELISA RIA - LC-MS/MS	- Active form of vitamin D: used to assess calcium metabolism - Evaluates disorders: hyperparathyroidism, renal failure, and vitamin D-dependent rickets
25-hydroxy vitamin D	- Chemiluminescent Immunoassay (CLIA) - LC-MS/MS	- Best marker for assessing vitamin D status - Used in diagnosing vitamin D deficiency, osteomalacia, and monitoring vitamin D supplementation
C- and N-telopeptides	- ELISA - Immunoassay automated analyzers	- Markers of bone resorption - Used to monitor osteoporosis therapy and evaluate bone turnover in metabolic bone diseases

(Table 3) cont.....

Test	Analytical Methods	Applied Aspects
Calcium – Total and Ionized	- Colorimetric method - Ion-selective electrode for ionized calcium	- Total calcium assesses overall calcium status - Ionized calcium is more accurate in evaluating true calcium levels in conditions like hyperparathyroidism, hypocalcemia
Magnesium	- Colorimetric method - Atomic absorption spectrophotometry	- Assesses magnesium deficiency or excess in conditions like chronic kidney disease, hypomagnesemia, and cardiovascular diseases
Osteocalcin	- ELISA - Radioimmunoassay (RIA)	- A marker of bone formation - Used in evaluating bone metabolism disorders, monitoring osteoporosis treatment, and assessing bone turnover
Phosphate	- Colorimetric method - Molybdate UV method	- Important for diagnosing and monitoring conditions like chronic kidney disease, hypophosphatemia, and bone disorders
PTH (Parathyroid Hormone)	- Chemiluminescent Immunoassay (CLIA) - ELISA	- Assesses parathyroid function in hyperparathyroidism and hypoparathyroidism - Evaluates calcium/phosphate imbalances
PTHrP (Parathyroid Hormone-related Peptide)	- Immunoassay - ELISA	- Used to differentiate hypercalcemia of malignancy from primary hyperparathyroidism - Monitors cancer-associated hypercalcemia
Pyridinolines	- HPLC - Immunoassay	- Markers of collagen breakdown and bone resorption - Useful in diagnosing and monitoring osteoporosis, Paget's disease, and metabolic bone disorders
Total and Bone-Specific ALP (Alkaline Phosphatase)	- Colorimetric method - Immunoassay for bone-specific isoform	- Total ALP assesses liver and bone diseases - Bone-specific ALP is used to monitor bone formation and osteoblastic activity in conditions like Paget's disease, rickets, and healing fractures

SUMMARY

Calcium and magnesium are crucial for bone health, neuromuscular function, and metabolism, and are regulated by hormones, namely, PTH and vitamin D. Imbalances can lead to conditions such as hypocalcemia, hypercalcemia, hypomagnesemia, and bone disorders like osteoporosis and osteomalacia. Lab tests for these minerals and hormones help diagnose these issues.

EXERCISE

- What is the primary role of parathyroid hormone in calcium regulation?
- Describe the clinical consequences of hypermagnesemia.
- What are the laboratory markers of bone formation, and how are they clinically significant?

- How is primary hyperparathyroidism differentiated from secondary hyperparathyroidism based on laboratory findings?
- What are the treatment options for osteoporosis, and how do they work?

CONCLUSION

Proper regulation of calcium and magnesium is crucial for neuromuscular and skeletal health. Parathyroid hormone and vitamin D play a central role in calcium homeostasis. Accurate diagnosis and lab assessments of minerals and bone markers are essential for effective treatment. Advancements in this area will improve patient outcomes in mineral metabolism and bone health.

ACKNOWLEDGEMENT AND CONFLICT OF INTEREST

No potential conflict of interest is declared by the author. It is also declared that the complete work is an individual effort by the author and that no financial/ administrative/ academic support was availed from any individual/ institution, or organization.

REFERENCES

[1] Saternus, R.; Vogt, T.; Reichrath, J. A Critical Appraisal of Strategies to Optimize Vitamin D Status in Germany, a Population with a Western Diet. *Nutrients,* **2019**, *11*(11), 2682. [http://dx.doi.org/10.3390/nu11112682] [PMID: 31698703]

[2] Wimalawansa, S.J.; Razzaque, M.S.; Al-Daghri, N.M. Calcium and vitamin D in human health: Hype or real? *J. Steroid Biochem. Mol. Biol.,* **2018**, *180*, 4-14. [http://dx.doi.org/10.1016/j.jsbmb.2017.12.009] [PMID: 29258769]

[3] Reid, I.R. Osteoporosis and Other Bone Diseases. *Lancet,* **2021**, *397*(10272), 1502-1512.

[4] Gragossian, A, Bashir K, Bhutta BS. Hypomagnesemia. [Updated 2023 Nov 30]. In: StatPearls [Internet]. Treasure Island (FL): StatPearls Publishing. **2025**. Available from: https://www.ncbi.nlm.nih.gov/books/NBK500003/

[5] Afroditi Roumpou, Andrea Palermo, Symeon Tournis, Valeria Hasenmajer, Janice L Pasieka, Gregory Kaltsas, Andrea Isidori, Eva Kassi, Bone in Parathyroid Diseases Revisited: Evidence From Epidemiological, Surgical and New Drug Outcomes, Endocrine Reviews, Volume 46, Issue 4, August **2025**, Pages 576–620. [http://dx.doi.org/10.1210/endrev/bnaf010]

CHAPTER 16

Musculoskeletal Diseases

Abstract: Musculoskeletal diseases impact bone metabolism and function. This chapter discusses the role of biochemistry in diagnosing and managing these diseases, emphasizing the importance of early diagnosis through biochemical testing to prevent severe complications such as fractures or deformities.

LEARNING OBJECTIVES

After reading this chapter, one should be able to:

- Discuss muscle function and the use and limitations of autoimmune testing in diagnosis.
- Discuss the causes, clinical signs, and symptoms of musculoskeletal disorders
- Describe how laboratory investigations are important in their detection, diagnosis, and management.
- Describe the specific laboratory investigations used in the management of musculoskeletal diseases.
- Recognize the analytical methods available for their measurement.
- Select laboratory investigations and interpret the analytical results in the context of the clinical signs and symptoms.

Simmi Kharb

KEY CONCEPTS: APPLIED ASPECTS OF MUSCLE FUNCTION

Muscle Function and the Use of Autoimmune Testing in Diagnosis (Table 1,2) [1 - 4]

Table 1. Muscle functions and applied aspects.

Aspect	Details
Key Concepts	-**Muscle Types**: Skeletal, cardiac, and smooth muscle -**Muscle Contraction**: Sliding filament theory involving actin and myosin -**Energy Sources**: ATP, creatine phosphate, glycolysis, and oxidative phosphorylation -**Neuromuscular Junction**: Transmission of nerve impulses to muscles *via* the neurotransmitter acetylcholine -**Muscle Fiber Types**: Slow-twitch (Type I) and fast-twitch (Type II) fibers.
Applied Aspects	
Muscle Enzymes	Creatine Kinase (CK): Key enzyme released during muscle damage *e.g.*, in rhabdomyolysis, muscular dystrophy CK levels are critical for diagnosing muscle injury or diseases. LDH and AST also rise in muscle damage, but are less specific.
Diagnostic Tests	Electromyography (EMG), muscle biopsy, and MRI to evaluate muscle function and structure. Biochemical markers (*e.g.*, CK, aldolase) complement these tests for muscle disease diagnosis.
Applications	-**Exercise Physiology**: Understanding muscle function helps in designing training programs. -**Rehabilitation**: Muscle function knowledge is crucial in physical therapy. -**Sports Medicine**: Enhancing muscle performance and recovery. -**Clinical Diagnostics**: Identifying muscle dysfunction in various diseases.

Table 2. Autoimmune testing in the diagnosis of musculoskeletal diseases.

Key Concept	Diagnostic use	Limitations
Antinuclear Antibodies (ANA)	Screening test for systemic autoimmune diseases like Systemic Lupus Erythematosus (SLE), Sjogren's syndrome, and mixed connective tissue diseases.	High sensitivity but low specificity; positive results in healthy individuals, infections, or other non-autoimmune conditions.
Anti-double-stranded DNA (Anti-dsDNA)	Specific marker for SLE, particularly for assessing disease activity and lupus nephritis.	False positives may occur in chronic infections and other autoimmune diseases.
Anti-Ro (SSA) and Anti-La (SSB) Antibodies	Diagnosis of Sjögren's syndrome and neonatal lupus; also seen in SLE.	Can be present in other autoimmune conditions, leading to diagnostic confusion.

(Table 2) cont.....

Key Concept	Diagnostic use	Limitations
Anti-Sm Antibodies	Highly specific for SLE; considered a confirmatory test.	Low sensitivity; found in only about 20-30% of SLE patients, thus may miss cases.
Anti-cyclic citrullinated peptide (Anti-CCP)	High specificity for Rheumatoid Arthritis (RA); helps in early diagnosis and predicting disease severity.	Can be negative in early stages of RA or in seronegative RA cases.
Rheumatoid Factor (RF)	Used in the diagnosis of RA, but can also be positive in other autoimmune conditions, infections, and aging.	Low specificity; can be positive in non-autoimmune diseases like chronic infections or in healthy elderly individuals.
Anti-neutrophil cytoplasmic antibodies (ANCA)	Useful in diagnosing vasculitis (*e.g.*, granulomatosis with polyangiitis).	Limited by false positives in infections, drug-induced vasculitides, and other inflammatory conditions.
Anti-thyroid peroxidase (Anti-TPO) and Anti-thyroglobulin (Anti-TG)	Markers for autoimmune thyroid diseases like Hashimoto's thyroiditis and Graves' disease.	Antibodies may be present in euthyroid individuals or those with non-autoimmune thyroid conditions.
Anti-Glutamic Acid Decarboxylase (GAD) Antibodies	Diagnostic for Type 1 diabetes and some neurological conditions (*e.g.*, stiff-person syndrome).	Can be found in other autoimmune conditions and healthy individuals, reducing specificity.
Anti-Phospholipid Antibodies (APA)	Helps diagnose antiphospholipid syndrome, which is associated with thrombosis and pregnancy complications.	Can be transiently positive in infections or drug reactions, requiring repeat testing for confirmation.
Anti-Smooth Muscle Antibodies (ASMA)	Used in the diagnosis of autoimmune hepatitis.	Low specificity; may be positive in viral hepatitis or other liver diseases.
Anti-Mitochondrial Antibodies (AMA)	Highly specific for Primary Biliary Cholangitis (PBC).	Low sensitivity; not present in all cases of PBC, potentially missing some diagnoses.
Anti-parietal Cell Antibodies	Associated with autoimmune gastritis and pernicious anemia.	Positive results are seen in other conditions, including aging or in individuals without clinical disease.

KEY CONCEPTS: USE AND LIMITATIONS OF AUTOIMMUNE TESTING IN DIAGNOSIS

Disease

Biochemical Basis, Relevant Lab Tests, Interpretation of Results

Laboratory Investigations for Detection, Diagnosis, and Management of Musculoskeletal Diseases (Tables 3 - 8) [1 - 4]

Table 3. Laboratory investigations for detection, diagnosis, and management of: ***Duchenne / becker dystrophy..***

Aspect	Details
Biochemical Basis	-**Duchenne Muscular Dystrophy (DMD)**: Absence of dystrophin protein due to mutations in the DMD gene. -**Becker Muscular Dystrophy (BMD)**: Reduced or abnormal dystrophin due to mutations in the same gene.
Lab Tests	-**Genetic Testing**: Identifies mutations in the DMD gene. -**Creatine Kinase (CK)**: Elevated levels, indicating muscle damage. -**Muscle Biopsy**: Shows absence or reduced dystrophin.
Interpretation	-**Genetic Testing**: Confirms diagnosis by identifying specific mutations .- **CK Levels**: Extremely high in DMD; moderately elevated in BMD. -**Muscle Biopsy**: Absent dystrophin in DMD; reduced in BMD.

Table 4. Laboratory investigations for detection, diagnosis, and management of: Osteoarthritis.

Aspect	Details
Biochemical Basis	-**Pathophysiology**: Degeneration of cartilage and changes in subchondral bone. -**Inflammatory Mediators**: Involvement of cytokines and enzymes like matrix metalloproteinases.
Lab Tests	-**X-rays**: Shows joint space narrowing, osteophytes. -**MRI**: Detailed imaging of cartilage and bone changes. -**Synovial Fluid Analysis**: Rule out other forms of arthritis.
Interpretation	-**Imaging**: Confirms osteoarthritis with characteristic joint changes. -**Synovial Fluid**: Typically non-inflammatory.

Table 5. Laboratory investigations for detection, diagnosis and management of: Rhabdomyolysis.

Aspect	Details
Biochemical Basis	-**Muscle Breakdown**: Release of muscle cell contents into the bloodstream. -**Myoglobin**: Released from damaged muscles, causing kidney damage.
Lab Tests	-**Creatine Kinase (CK)**: Markedly elevated levels. -**Myoglobin**: Elevated in blood and urine. -**Electrolytes**: Abnormalities like hyperkalemia. -**Renal Function Tests**: Elevated creatinine and BUN.
Interpretation	-**CK Levels**: Extremely high, often > 10,000 IU/L. -**Myoglobin**: Presence in urine can cause dark-colored urine. -**Electrolytes and Renal Function**: Monitor for kidney damage and electrolyte imbalances.

Table 6. Laboratory investigations for detection, diagnosis, and management of: Rheumatoid Arthritis.

Aspect	Details
Biochemical Basis	-**Autoimmunity**: Chronic inflammation of synovial joints due to the immune system attacking the synovium. -**Cytokines**: Role of TNF-α, IL-1, and IL-6 in inflammation and joint damage.
Lab Tests	-**Rheumatoid Factor (RF)**: Often elevated. -**Anti-Cyclic Citrullinated Peptide (Anti-CCP) Antibodies**: High specificity for RA. -**ESR and CRP**: Elevated during active disease. -**Joint Imaging**: X-ray, MRI, and ultrasound to assess joint damage.
Interpretation	-**RF and Anti-CCP**: Presence supports the diagnosis of RA. -**ESR and CRP**: Indicators of inflammation. -**Imaging**: Shows joint erosion and other changes characteristic of RA.

Table 7. Laboratory investigations for detection, diagnosis, and management of: Systemic Lupus Erythematosus (SLE).

Aspect	Details
Biochemical Basis	-**Autoimmunity**: The immune system attacks multiple organs and tissues. -**Autoantibodies**: Production of antibodies against nuclear components (*e.g.*, anti-dsDNA, anti-Smith).
Lab Tests	-**ANA (Antinuclear Antibodies)**: Screening test, usually positive. -**Anti-dsDNA and Anti-Smith Antibodies**: Highly specific for SLE. -**Complement Levels (C3, C4)**: Often decreased during active disease. -**ESR and CRP**: Elevated during flares.
Interpretation	-**ANA**: High sensitivity but low specificity .- **Anti-dsDNA and Anti-Smith**: High specificity for SLE. -**Complement Levels**: Low during active disease. -**ESR and CRP**: Elevated in active disease.

Table 8. Laboratory investigations for detection, diagnosis, and management of: Vasculitis.

Aspect	Details
Biochemical Basis	-**Inflammation of Blood Vessels**: Leads to vessel wall damage and organ ischemia. -**Immune-Mediated**: Often involves autoantibodies (*e.g.*, ANCA in granulomatosis with polyangiitis).
Lab Tests	-**ANCA (Anti-Neutrophil Cytoplasmic Antibodies)**: c-ANCA and p-ANCA are specific for certain types of vasculitides. -**ESR and CRP**: Elevated levels, indicating inflammation. -**Biopsy**: Confirms diagnosis by showing vessel inflammation and damage. -**Imaging**: Angiography, MRI, CT to assess the extent and location of vessel involvement.
Interpretation	-**ANCA**: Helps in diagnosing specific types of vasculitides. -**ESR and CRP**: Indicators of systemic inflammation. -**Biopsy**: Definitive diagnosis showing vasculitis.

Relevant Laboratory Tests, Analytics Methods Available, and Applied Aspects for

Analytical Methods Available for Laboratory Investigations and Interpretation of Analytical Results (Tables 9 - 14) [1 - 4]

Table 9. Relevant laboratory tests, analytics methods available, and applied aspects for: Anti-CCP antibodies.

Aspect	Details
Biochemical Basis	-**Autoantibodies** targeting cyclic citrullinated peptides, which are associated with Rheumatoid Arthritis (RA).
Analytical Methods	-**ELISA (Enzyme-Linked Immunosorbent Assay)**: Most commonly used method. -**Chemiluminescent Immunoassay**: Also used for detection.
Clinical Use	-**Diagnosis**: Highly specific for RA, often used in conjunction with Rheumatoid Factor (RF). -**Prognosis**: Associated with more severe disease and joint damage.
Interpretation	-**Positive Result**: Strongly suggests RA, especially if RF is also positive. -**Negative Result**: Does not rule out RA but may indicate a milder form.
Applied Aspects	-**Early Diagnosis**: Helps in the early identification of RA, leading to timely initiation of treatment. -**Disease Management**: Monitoring response to therapy.

Table 10. Relevant laboratory tests, analytics methods available, and applied aspects for: Anti-ds DNA.

Aspect	Details
Biochemical Basis	-**Autoantibodies** against double-stranded DNA, associated with Systemic Lupus Erythematosus (SLE).
Analytical Methods	-**ELISA**: Common method for initial screening. -**Farr Assay**: Radioimmunoassay, highly specific. -**Indirect Immunofluorescence**: Crithidia luciliae assay.
Clinical Use	-**Diagnosis**: Specific for SLE, used to confirm diagnosis. -**Disease Activity**: Levels correlate with disease activity, particularly renal involvement.
Interpretation	-**Positive Result**: Strongly indicates SLE. -**High Levels**: Suggest active disease, especially lupus nephritis. -**Low Levels**: May indicate remission.
Applied Aspects	-**Monitoring**: Regular testing to assess disease activity and treatment response. -**Management**: Adjusting therapy based on antibody levels and clinical symptoms.

Table 11. Relevant laboratory tests, analytics methods available, and applied aspects for: Anti-neutrophilic Cytoplasmic Antibodies (ANCA).

Aspect	Details
Biochemical Basis	-**Autoantibodies** against components of neutrophil cytoplasm, associated with vasculitis.
Analytical Methods	-**Indirect Immunofluorescence**: Screening test to detect c-ANCA and p-ANCA. -**ELISA**: To confirm and specify the target antigen (*e.g.*, PR3 for c-ANCA, MPO for p-ANCA).
Clinical Use	-**Diagnosis**: Helps diagnose types of vasculitides such as granulomatosis with polyangiitis (c-ANCA) and microscopic polyangiitis (p-ANCA).
Interpretation	-**Positive c-ANCA**: Indicates granulomatosis with polyangiitis. -**Positive p-ANCA**: Suggests microscopic polyangiitis or other forms of vasculitis.
Applied Aspects	-**Diagnosis**: Essential for identifying and differentiating vasculitic syndromes. -**Monitoring**: Used to track disease activity and response to treatment.

Table 12. Relevant laboratory tests, analytics methods available, and applied aspects for: Anti-nuclear Antibodies (ANA and Specific Antibodies SSA, SSB, Sm, RNP).

Aspect	Details
Biochemical Basis	-**Autoantibodies** against nuclear components, associated with various autoimmune diseases.
Analytical Methods	-**Indirect Immunofluorescence (IF)**: Standard screening test. -**ELISA**: For specific antibodies such as SSA (Ro), SSB (La), Sm, and RNP.
Clinical Use	-**Screening**: ANA is a broad screening test for autoimmune diseases. -**Diagnosis**: Specific antibodies (SSA, SSB, Sm, RNP) help diagnose SLE, Sjögren's syndrome, and Mixed Connective Tissue Disease (MCTD).
Interpretation	-**Positive ANA**: Suggests autoimmune disease but requires further specific testing. -**Specific Antibodies**: Confirm diagnosis of specific autoimmune diseases.
Applied Aspects	-**Screening**: Initial step in evaluating patients with suspected autoimmune diseases. -**Diagnosis**: Confirming and differentiating specific autoimmune conditions.

Table 13. Relevant laboratory tests, analytics methods available, and applied aspects for: Rheumatoid Factor (RF).

Aspect	Details
Biochemical Basis	-**Autoantibodies** targeting the Fc region of IgG, associated with rheumatoid arthritis.
Analytical Methods	-**ELISA**: Common method for detection. -**Latex Agglutination Test**: Traditional method. -**Nephelometry and Turbidimetry**: Quantitative methods.
Clinical Use	-**Diagnosis**: Used in diagnosing RA, but also seen in other autoimmune diseases and chronic infections.

(Table 13) cont.....

Aspect	Details
Interpretation	-**Positive Result**: Supports diagnosis of RA, especially if combined with clinical symptoms and other tests (*e.g.*, anti-CCP). -**Negative Result**: Does not exclude RA.
Applied Aspects	-**Diagnosis**: Part of the diagnostic criteria for RA. -**Prognosis**: High levels may indicate more severe disease. -**Monitoring**: Tracking disease progression and response to treatment.

Table 14. Relevant laboratory tests, analytics methods available, and applied aspects for: Serum Creatine Kinase (CK).

Aspect	Details
Biochemical Basis	-**Enzyme** is released into the bloodstream from damaged muscle tissue.
Analytical Methods	-**Spectrophotometric Assay**: Measures enzyme activity. -**Automated Chemistry Analyzers**: Common in clinical labs for routine CK measurement.
Clinical Use	-**Diagnosis**: Detects muscle damage in conditions like rhabdomyolysis, myocardial infarction, and myopathies (*e.g.*, Duchenne/Becker dystrophy).
Interpretation	-**Elevated Levels**: Indicate muscle damage; very high levels are seen in rhabdomyolysis. -**CK-MB Fraction**: Specific for cardiac muscle damage (*e.g.*, myocardial infarction).
Applied Aspects	-**Diagnosis**: Identifying muscle damage and differentiating between cardiac and skeletal muscle sources. -**Monitoring**: Tracking muscle disease progression and response to treatment. -**Management**: Guiding therapeutic interventions in conditions like rhabdomyolysis and muscular dystrophies.

SUMMARY

Musculoskeletal diseases involve biochemical imbalances affecting bone strength, mineralization, and turnover. Osteoporosis is characterized by reduced bone mass and increased fracture risk, and is diagnosed using bone density assessments and markers of bone turnover. Osteomalacia results from vitamin D deficiency, leading to impaired bone mineralization, which is diagnosed by low serum calcium and phosphate, and elevated PTH. Paget's disease involves increased bone remodelling, marked by elevated ALP. Inflammatory conditions show elevated inflammatory markers and abnormal bone turnover markers.

CONCLUSION

Biochemical markers are crucial for early diagnosis and monitoring of musculoskeletal diseases. Clinicians need to combine these markers with clinical findings and imaging studies for effective management. Understanding calcium-phosphate metabolism, vitamin D status, and bone turnover markers helps optimize therapeutic approaches. Ongoing advancements in biomarker discovery and analytical methods improve the accuracy of diagnosis and treatment monitoring.

EXERCISES

- What are the primary biochemical markers used to diagnose osteoporosis?
- Explain the role of vitamin D in bone metabolism and how its deficiency can lead to osteomalacia.
- How is Paget's disease biochemically diagnosed and monitored?
- Discuss the use of C-terminal Telopeptide (CTX) in monitoring bone resorption in patients with osteoporosis.
- How do inflammatory markers correlate with musculoskeletal diseases like rheumatoid arthritis?
- Describe the role of Alkaline Phosphatase (ALP) in diagnosing and managing bone-related diseases.

ACKNOWLEDGEMENT AND CONFLICT OF INTEREST

No potential conflict of interest is declared by the author. It is also declared that the complete work is an individual effort by the author and that no financial/ administrative/ academic support was availed from any individual/ institution, or organization.

REFERENCES

[1] Kaur J, Khare S, Givler A. Vitamin D Deficiency. [Updated 2025 Feb 15]. In: StatPearls [Internet]. Treasure Island (FL): StatPearls Publishing; **2026** Jan.
Available from: https://www.ncbi.nlm.nih.gov/books/NBK532266/

[2] Mobasheri, A, Thudium CS, Bay-Jensen AC, Maleitzke T, Geissler S, Duda GN, Winkler T. Biomarkers for osteoarthritis: Current status and future prospects. *Best Pract Res Clin Rheumatol.*, **2023**, *37*, 101852.
[http://dx.doi.org/10.1016/j.berh.2023.101852]

[3] Schini, M.; Vilaca, T.; Gossiel, F.; Salam, S.; Eastell, R.; Markers, B.T. Bone Turnover Markers: Basic Biology to Clinical Applications. *Endocr. Rev.*, **2023**, *44*(3), 417-473.
[http://dx.doi.org/10.1210/endrev/bnac031] [PMID: 36510335]

[4] Gennari, L.; Rendina, D.; Merlotti, D.; Cavati, G.; Mingiano, C.; Cosso, R.; Materozzi, M.; Pirrotta, F.; Abate, V.; Calabrese, M.; Falchetti, A. Update on the pathogenesis and genetics of Paget's disease of bone. *Front Cell Dev Biol,* **2022**, *10*, 932065.
[http://dx.doi.org/10.3389/fcell.2022.932065]

CHAPTER 17

Neurological and Psychiatric Disorders

Abstract: Neurological and psychiatric disorders are complex conditions that involve the interplay of various biochemical pathways. This study guide focuses on the applied aspects of clinical chemistry in diagnosing and managing these disorders. It covers the biochemical markers associated with different neurological and psychiatric conditions, the laboratory techniques used to measure these markers, and the interpretation of results in a clinical context. The guide also highlights the importance of a multidisciplinary approach and ethical considerations in the biochemical assessment of these disorders.

LEARNING OBJECTIVES

After reading this chapter, one should be able to:

- Discuss the biochemical basis and clinical manifestations of neurological and psychiatric disorders.
- Discuss the causes, clinical signs, and symptoms of the following disorders, some of which are primary neurological disorders and some of which have neurological manifestations.
- Describe which laboratory investigations are important in their detection, diagnosis, and management.
- Discuss the following specific laboratory investigations related to the study of neurological disorders.
- Recognise the analytical methods available for their measurement.
- Select laboratory investigations and interpret the analytical results in the context of the clinical signs and symptoms.

Simmi Kharb

KEY CONCEPTS

Biochemical Basis and Clinical Manifestations of Neurological and Psychiatric Disorders (Tables 1 and 2) [1 - 3]

Table 1. Blood-Brain Barrier (BBB).

Aspect	Details
Key Concepts	-**Structure**: The BBB is formed by endothelial cells of brain capillaries, tightly joined by tight junctions, along with astrocyte end-feet and pericytes **Function**: The BBB regulates the passage of substances from the bloodstream into the brain, protecting the CNS from toxins and pathogens while allowing essential nutrients to pass
Physiological Roles	-**Protection**: Shields the brain from harmful substances and pathogens -**Homeostasis**: Maintains a stable environment for neuronal function -**Selective Permeability**: Allows the passage of essential molecules like glucose and amino acids while restricting harmful substances
Pathophysiology	-**Disruption**: Conditions like multiple sclerosis, stroke, trauma, and infections can disrupt the BBB, leading to increased permeability and potential neuroinflammation -**Inflammation and Autoimmunity**: Inflammatory cytokines can weaken BBB integrity, facilitating the entry of immune cells into the CNS
Clinical Applications	-**Drug Delivery**: Challenges in delivering therapeutic agents to the brain due to the restrictive nature of the BBB -**Biomarkers**: BBB integrity can be assessed through imaging techniques and Cerebrospinal Fluid (CSF) analysis -**Therapies**: Research into methods to transiently open the BBB for drug delivery, such as focused ultrasound and chemical agents.
Research Directions	-**Nanotechnology**: Development of nanoparticles to cross the BBB for targeted drug delivery -**Molecular Mechanisms**: Understanding the molecular pathways regulating BBB function to develop strategies to protect or restore its integrity -**Neurodegenerative Diseases**: Investigating the role of BBB dysfunction in diseases like Alzheimer's and Parkinson's.

Table 2. Paraneoplastic syndromes [1, 2].

Aspect	Details
Key Concepts	-**Definition**: Paraneoplastic syndromes are a group of disorders that occur in patients with cancer, where symptoms are not directly related to tumor invasion or metastasis but are caused by immune-mediated mechanisms or hormone-like substances produced by the tumor -**Mechanism**: Often involves immune cross-reactivity where the body's immune response against tumor antigens also targets similar antigens in the nervous system

(Table 3) cont.....

Aspect	Details
Common Syndromes	-**Neurological Paraneoplastic Syndromes**: Include Lambert-Eaton myasthenic syndrome, paraneoplastic cerebellar degeneration, opsoclonus-myoclonus syndrome, and limbic encephalitis -**Symptoms**: Can vary widely but often include neurological deficits such as ataxia, myoclonus, encephalopathy, neuropathy, and autonomic dysfunction
Pathophysiology	-**Autoantibodies**: Production of antibodies against neuronal antigens (*e.g.*, Hu, Yo, Ri, Ma2) that are also expressed by the tumor -**Immune Response**: Immune cells, such as T cells, may also play a role in the damage to the nervous system
Diagnostic Tests	-**Autoantibody Testing**: Detection of specific paraneoplastic antibodies (*e.g.*, anti-Hu, anti-Yo, anti-Ri) in serum or CSF -**Imaging**: MRI may show characteristic changes depending on the affected CNS region -**Tumor Screening**: Comprehensive screening to identify the underlying malignancy, including CT, PET scans, and tumor markers
Clinical Management	-**Treating the Underlying Cancer**: Often the primary approach, as successful treatment of the cancer can lead to improvement of neurological symptoms -**Immunotherapy**: Use of corticosteroids, IVIG, plasmapheresis, and other immunosuppressive therapies to reduce the autoimmune response -**Symptomatic Treatment**: Management of specific symptoms and supportive care
Research Directions	-**Immunopathogenesis**: Further understanding of the immune mechanisms leading to paraneoplastic syndromes -**Biomarkers**: Identification of new biomarkers for early detection and better diagnosis -**Therapeutic Strategies**: Development of targeted therapies to modulate the immune response without compromising anti-tumor immunity

Biochemical Basis, Relevant Lab Tests, Interpretation, and Applied Aspects

Biochemical Basis, Relevant Lab Tests, Interpretation, and Applied Aspects of Neurological and Psychiatric Disorders (Tables 3-8)

Table 3. Acute porphyrias.

Aspect	Details
Biochemical Basis	-**Metabolic Disorders**: Acute porphyrias are caused by defects in heme biosynthesis, leading to accumulation of porphyrin precursors.
Types	-**Acute Intermittent Porphyria (AIP), Variegate Porphyria (VP), Hereditary Coproporphyria (HCP)**.
Symptoms	-**Acute Attacks**: Abdominal pain, neurological symptoms, and psychiatric manifestations.

(Table 3) cont.....

Aspect	Details
Relevant Lab Tests	-**Urine Tests**: Elevated levels of Porphobilinogen (PBG) and Aminolevulinic Acid (ALA). -**Stool Tests**: Elevated coproporphyrin and protoporphyrin (for VP, HCP).
Interpretation	-**Positive Results**: High PBG and ALA in urine during an acute attack confirm diagnosis. -**Stool Porphyrins**: Help distinguish between different types of porphyria.
Applied Aspects	-**Management**: Avoidance of triggering factors, administration of heme arginate or glucose during attacks. -**Genetic Counseling**: Screening of family members.

Table 4. Alzheimer's disease.

Aspect	Details
Biochemical Basis	-**Amyloid Plaques**: Accumulation of beta-amyloid peptides. -**Neurofibrillary Tangles**: Hyperphosphorylated tau protein.
Symptoms	-**Cognitive Decline**: Memory loss, disorientation, language difficulties, and impaired judgment.
Relevant Lab Tests	-**CSF Biomarkers**: Decreased beta-amyloid 1-42, increased total tau, and phosphorylated tau. -**Genetic Tests**: APOE ε4 allele testing.
Interpretation	-**CSF Biomarkers**: Patterns of decreased amyloid and increased tau are indicative of Alzheimer's -**Genetic Testing**: APOE ε4 allele increases risk but is not diagnostic.
Applied Aspects	-**Diagnosis**: Early diagnosis with CSF biomarkers and imaging (MRI, PET) -**Management**: Symptomatic treatment, cholinesterase inhibitors, and NMDA receptor antagonists.

Table 5. Meningitis.

Aspect	Details
Biochemical Basis	-**Infection**: Caused by bacteria, viruses, fungi, or parasites, leading to inflammation of the meninges.
Symptoms	-**Acute Onset**: Headache, fever, neck stiffness, altered mental status.
Relevant Lab Tests	-**CSF Analysis**: Elevated white cell count, increased protein, decreased glucose (bacterial), normal glucose (viral). -**Microbiology**: Gram stain, culture, and PCR for specific pathogens.
Interpretation	-**CSF Results**: Elevated white cell count and protein with decreased glucose suggest bacterial meningitis; normal glucose suggests viral meningitis. -**Microbiology**: Identifies causative organism.

(Table 5) cont.....

Aspect	Details
Applied Aspects	-**Treatment**: Empiric antibiotics for bacterial meningitis, antivirals for viral meningitis. -**Prevention**: Vaccination, prophylactic antibiotics for close contacts.

Table 6. Multiple sclerosis.

Aspect	Details
Biochemical Basis	-**Autoimmune Disorder**: Immune-mediated attack on myelin sheaths in the central nervous system.
Symptoms	-**Relapsing-Remitting Course**: Episodes of neurological dysfunction followed by recovery -**Progressive Course**: Gradual accumulation of disability.
Relevant Lab Tests	-**MRI**: White matter lesions (plaques) in the CNS **CSF Analysis**: Presence of oligoclonal bands and increased IgG index.
Interpretation	-**MRI**: Characteristic demyelinating lesions in the brain and spinal cord. -**CSF**: Oligoclonal bands support diagnosis but are not specific.
Applied Aspects	-**Treatment**: Immunomodulatory therapies (*e.g.*, interferons, glatiramer acetate), monoclonal antibodies (*e.g.*, natalizumab), symptomatic treatment. -**Monitoring**: Regular MRI and clinical assessment to monitor disease progression and treatment response.

Table 7. Myasthenia gravis.

Aspect	Details
Biochemical Basis	-**Autoimmune Disorder**: Autoantibodies against Acetylcholine Receptors (AChR) at the neuromuscular junction.
Symptoms	-**Muscle Weakness**: Fluctuating muscle weakness, particularly affecting ocular, bulbar, and limb muscles.
Relevant Lab Tests	-**Serological Tests**: Anti-AChR antibodies, anti-MuSK (Muscle-specific Kinase) antibodies -**Electrophysiological Tests**: Repetitive nerve stimulation, single-fiber Electromyography (EMG).
Interpretation	-**Positive Antibodies**: Presence of anti-AChR or anti-MuSK antibodies confirms diagnosis. -**EMG**: Decremental response on repetitive nerve stimulation.
Applied Aspects	-**Treatment**: Acetylcholinesterase inhibitors (*e.g.*, pyridostigmine), immunosuppressants (*e.g.*, corticosteroids, azathioprine), thymectomy in selected cases. -**Monitoring**: Regular assessment of muscle strength and function.

Table 8. Paraneoplastic syndromes.

Aspect	Details
Biochemical Basis	**-Immune-Mediated**: Autoantibodies produced in response to tumor antigens cross-react with neuronal antigens.
Symptoms	**-Neurological Deficits**: Variable depending on the affected part of the nervous system; can include limbic encephalitis, cerebellar degeneration, and myelopathy.
Relevant Lab Tests	**-Serum and CSF Antibodies**: Anti-Hu, anti-Yo, anti-Ri, anti-Ma2, and other paraneoplastic antibodies. **-Imaging**: MRI to detect neurological involvement. **-Tumor Screening**: CT, PET scans to identify the underlying malignancy.
Interpretation	**-Positive Antibodies**: Presence of specific paraneoplastic antibodies supports diagnosis. **-Imaging**: Identifies neurological involvement and underlying malignancy.
Applied Aspects	**-Treatment**: Treat underlying cancer, immunotherapy (*e.g.*, corticosteroids, IVIG, plasmapheresis), and symptomatic management **-Monitoring**: Regular follow-up for cancer recurrence and neurological symptoms.

Laboratory Investigations

Laboratory Investigations in their Detection, Diagnosis, and Management and Interpretation (Tables 9 - 19) [4]

Table 9. Interpretation of analytical results: Aminolaevulinic Acid (ALA).

Aspect	Details
Relevance	Diagnostic marker for acute porphyrias.
Sample Type	Urine.
Analytical Methods	**-High-Performance Liquid Chromatography (HPLC)** **-Spectrophotometric Assay** **-Mass Spectrometry (MS)**.
Interpretation	Elevated levels indicate acute porphyric attacks.
Applied Aspects	Monitor during and between attacks for disease management.

Table 10. Interpretation of analytical results: Anti-Acetylcholine Receptor Antibodies..

Aspect	Details
Relevance	Diagnostic marker for Myasthenia Gravis (MG).
Sample Type	Serum.
Analytical Methods	**-Radioimmunoassay (RIA)** **-Enzyme-Linked Immunosorbent Assay (ELISA)**.

(Table 10) cont.....

Aspect	Details
Interpretation	Presence confirms MG diagnosis; higher levels correlate with disease severity.
Applied Aspects	Guide diagnosis and monitor response to immunosuppressive therapy.

Table 11. Interpretation of analytical results: Anti-Hu, Anti-Yo Antibodies..

Aspect	Details
Relevance	Diagnostic markers for paraneoplastic syndromes.
Sample Type	Serum, Cerebrospinal Fluid (CSF).
Analytical Methods	**-Immunohistochemistry (IHC)** **-Western Blotting** **-ELISA** **-Indirect Immunofluorescence (IIF)**.
Interpretation	Positive results indicate paraneoplastic syndromes associated with specific cancers.
Applied Aspects	Guide cancer screening and neurological evaluation.

Table 12. Interpretation of analytical results: CSF Beta-2-Transferrin in Rhinorrhea and Otorrhea (Fistula).

Aspect	Details
Relevance	Specific marker for Cerebrospinal Fluid (CSF) leakage.
Sample Type	Nasal or ear discharge.
Analytical Methods	**-Immunofixation Electrophoresis (IFE)** **-Western Blotting** **-Mass Spectrometry (MS)**.
Interpretation	Presence confirms CSF leakage, aiding in the diagnosis of fistulas.
Applied Aspects	Guide the surgical repair and management of CSF leaks.

Table 13. Interpretation of analytical results: CSF Glucose.

Aspect	Details
Relevance	An indicator of Central Nervous System (CNS) infections, such as bacterial meningitis.
Sample Type	CSF.
Analytical Methods	**-Glucose Oxidase Method** **-Hexokinase Method**.
Interpretation	Low CSF glucose relative to blood glucose suggests bacterial meningitis or other infectious/inflammatory conditions.
Applied Aspects	Differentiate between bacterial and viral meningitis, guiding appropriate treatment.

Table 14. Interpretation of analytical results: CSF Protein.

Aspect	Details
Relevance	An indicator of CNS infections, inflammation, and demyelinating diseases.
Sample Type	CSF.
Analytical Methods	**-Biuret Method** **-Turbidimetric Method** **-Spectrophotometric Assay**.
Interpretation	Elevated levels indicate infection, inflammation, or other CNS pathology.
Applied Aspects	Aid in the diagnosis and monitoring of CNS diseases such as multiple sclerosis and meningitis.

Table 15. Interpretation of analytical results: Examination of Synovial Fluid.

Aspect	Details
Relevance	Diagnosis of joint disorders, including infections, crystal-induced arthritis, and inflammatory conditions.
Sample Type	Synovial fluid.
Analytical Methods	-**Microscopy** for cell count and crystals -**Gram Stain and Culture** for infection -**Biochemical Analysis** for glucose and protein.
Interpretation	Findings guide diagnosis: high WBCs indicate inflammation; crystals suggest gout or pseudogout; positive culture indicates infection.
Applied Aspects	Guide treatment of joint disorders, such as antibiotics for infections and anti-inflammatory therapy for crystal-induced arthritis.

Table 16. Interpretation of analytical results: Oligoclonal Banding (Isoelectric Focusing).

Aspect	Details
Relevance	Diagnostic marker for multiple sclerosis and other inflammatory CNS diseases.
Sample Type	CSF.
Analytical Methods	**-Isoelectric Focusing (IEF)** followed by **Immunoblotting**.
Interpretation	Presence of oligoclonal bands in CSF but not in serum supports the diagnosis of multiple sclerosis.
Applied Aspects	Confirm diagnosis of multiple sclerosis and monitor disease progression.

Table 17. Interpretation of analytical results: Paraneoplastic Antibodies.

Aspect	Details
Relevance	Indicators of paraneoplastic syndromes associated with cancer.
Sample Type	Serum, CSF.
Analytical Methods	**-Indirect Immunofluorescence (IIF)** **-Western Blotting** **-ELISA** **-Radioimmunoassay (RIA)**.
Interpretation	Positive results indicate paraneoplastic syndromes and guide cancer screening.
Applied Aspects	Guide the diagnosis and management of neurological symptoms related to paraneoplastic syndromes.

Table 18. Interpretation of analytical results: Porphobilinogen (PBG).

Aspect	Details
Relevance	Diagnostic marker for acute porphyrias.
Sample Type	Urine.
Analytical Methods	**-Watson-Schwartz Test** **-Hoesch Test** **-HPLC** **-Mass Spectrometry (MS)**.
Interpretation	Elevated PBG levels during acute attacks confirm the diagnosis of acute porphyria.
Applied Aspects	Monitor during acute attacks and between episodes for effective management.

Table 19. Interpretation of analytical results: Porphyrins (Urine, Feces, Serum).

Aspect	Details
Relevance	Diagnostic markers for various types of porphyrias.
Sample Type	Urine, feces, and serum.
Analytical Methods	**-Spectrophotometry** **-HPLC** **-Mass Spectrometry (MS)**.
Interpretation	Elevated levels indicate specific types of porphyria and help distinguish between acute and cutaneous forms.
Applied Aspects	Guide diagnosis and management of porphyrias, including dietary and lifestyle modifications and medication adjustments.

SUMMARY

Understanding the intricate web of biochemical pathways that govern brain function is essential for unravelling the complexities of neurological and psychiatric disorders. Specific neurochemical abnormalities have been associated with disorders like Alzheimer's disease, Parkinson's disease, schizophrenia, and depression. Diagnostic tools such as Cerebrospinal Fluid (CSF) analysis, blood tests, and imaging techniques are indispensable for identifying these conditions. This chapter delves into the latest advancements in neurochemical biomarker identification, highlighting their potential to revolutionize clinical practice and enhance patient outcomes. A deeper comprehension of these markers will refine diagnostic and therapeutic strategies, leading to more effective management of these disorders.

EXERCISES

- Which biomarkers are commonly associated with Alzheimer's disease?
- Discuss the test used for the diagnosis of multiple sclerosis.
- What is the role of serotonin in the pathophysiology of depression?
- Explain the mechanism of neurochemical markers in the early diagnosis of psychiatric disorders.

CONCLUSION

Biochemical markers are crucial for comprehending, diagnosing, and treating neurological and psychiatric disorders. Integrating clinical chemistry into the diagnostic process allows for identifying specific biomarkers linked to various neuropsychiatric conditions. Progress in molecular biology and neurochemistry offers hope for creating new diagnostic tools and treatments. Persistent research and implementation of these discoveries in clinical settings will enhance diagnostic precision, personalized treatment, and overall patient care.

REFERENCES

[1] Blennow, K.; Zetterberg, H. Biomarkers for Alzheimer's disease: current status and prospects for the future. *J. Intern. Med.*, **2018**, *284*(6), 643-663. [http://dx.doi.org/10.1111/joim.12816] [PMID: 30051512]

[2] Pietrzak, R.H.; Lim, Y.Y.; Ames, D.; Harrington, K.; Restrepo, C.; Martins, R.N.; Rembach, A.; Laws, S.M.; Masters, C.L.; Villemagne, V.L.; Rowe, C.C.; Maruff, P. Trajectories of memory decline in preclinical Alzheimer's disease: results from the Australian Imaging, Biomarkers and Lifestyle Flagship Study of Ageing. *Neurobiol. Aging,* **2015**, *36*(3), 1231-1238. [http://dx.doi.org/10.1016/j.neurobiolaging.2014.12.015] [PMID: 25585532]

[3] Rucker, J.T.; Lamichhane, B.; Haren, W.B.; Weinstein, B.L.; Madan, A. Sexually dimorphic association of Catechol-O-Methyltransferase (COMT) functional polymorphism with dimensional personality pathology: a preliminary study. *Brain Sciences,* **2025**, *15*(12), 1337. [http://dx.doi.org/10.3390/brainsci15121337]

[4] Hasler, G. Pathophysiology of depression: do we have any solid evidence of interest to clinicians? *World Psychiatry,* **2010**, *9*(3), 155-161.
[http://dx.doi.org/10.1002/j.2051-5545.2010.tb00298.x] [PMID: 20975857]

CHAPTER 18

Pediatric Clinical Chemistry

Abstract: Pediatric clinical chemistry involves specialized techniques and considerations due to the physiological differences between children and adults. This field plays a crucial role in diagnosing and managing various pediatric disorders, such as congenital hypothyroidism, cystic fibrosis, and type 1 diabetes mellitus. This chapter outlines the biochemical underpinnings of common pediatric diseases, discusses the appropriate laboratory tests and their interpretation, and addresses the challenges of using pediatric reference ranges. Understanding these aspects is essential for accurate diagnosis and effective management, ultimately improving patient outcomes.

Keywords: Clinical laboratory, Diagnosis, Management, Pediactric, Reference range.

LEARNING OBJECTIVES

After reading this chapter, one should be able to:

- Discuss special considerations in relation to the provision of a general pediatric clinical laboratory service.
- Discuss the causes, clinical signs, and symptoms of the conditions that may present in a general pediatric laboratory.
- Describe which laboratory investigations are important in their detection, diagnosis, and management.
- Discuss specific laboratory investigations for the evaluation of disorders presenting in the neonate and in childhood.
- Recognize the analytical methods available for their measurement.
- Select laboratory investigations and interpret the analytical results in the context of clinical signs and symptoms.
- Recognize the importance of blood volume limitations in method selection.
- Recognize that the population-based reference intervals for many chemistry tests change throughout childhood.
- Discuss the importance of and challenges associated with determining pediatric reference intervals.

Simmi Kharb

KEY CONCEPTS

Key Concepts in General Pediatric Clinical Laboratory Service

Special Considerations in Relation to the Provision of a General Pediatric Clinical Laboratory Service (Table 1)

Table 1. Concept of general pediatric clinical laboratory service.

Key Concept	Details
Collection of Heel Stick Samples	-**Definition:** A minimally invasive blood collection method typically used for newborns and infants. -**Procedure:** Involves making a small puncture in the infant's heel to collect a few drops of blood. -**Uses:** Common for newborn screening tests, glucose monitoring, and bilirubin levels.
Issues with Capillary Specimens	-**Contamination Risk:** Capillary samples are prone to contamination with interstitial and intracellular fluids. -**Hemolysis:** Increased risk due to the small volume and handling techniques. -**Inconsistent Results:** Variability in blood flow and sample collection technique can lead to inconsistencies.
Pediatric Reference Intervals	-**Definition:** Age-appropriate reference ranges for interpreting laboratory results in pediatric populations. -**Importance:** Essential for accurate diagnosis and treatment, as children have different normal ranges than adults. -**Dynamic Changes:** Influenced by growth, development, and puberty stages.
Dynamic Changes with Growth, Development, and Puberty	-**Metabolic Rate Variability:** Changes in metabolism, hormone levels, and organ function as children grow. -**Impact on Lab Values:** Parameters such as hemoglobin, liver enzymes, and hormone levels vary significantly with age and developmental stage. -**Monitoring:** Requires age-specific reference intervals.
Challenges with Determining Pediatric Reference Intervals	-**Limited Data:** Fewer studies available compared to adult populations. -**Ethical Issues:** Concerns with obtaining blood samples from healthy children. -**Variability:** Differences in age, gender, ethnicity, and nutritional status make standardization difficult. -**Cost:** High costs of establishing and validating intervals.

(Table 1) cont.....

Key Concept	Details
Sample Volume and Collection Issues	**-Volume Constraints:** Pediatric patients have limited blood volume, necessitating minimal sample sizes. **-Techniques:** Use of micro-collection tubes, capillary sticks, and minimizing the number of tests. **-Pain and Stress:** Techniques that reduce pain and stress are essential for compliance and accuracy.
Sweat Collection	**-Purpose:** Commonly used for cystic fibrosis screening by measuring chloride levels in sweat. **-Procedure:** Involves stimulating sweat glands (typically on the forearm) and collecting sweat using specialized collection devices. **-Challenges:** Small sample volume, variability in sweat production, and technical difficulties.

DISEASES

Biochemical Basis, Relevant Lab Tests, Interpretation of Results, and Applied Aspects for

Laboratory Investigations in Detection, Diagnosis and Management of Disorders Presenting in the Neonate and in Childhood (Table 2) [1 - 4]

Table 2. Biochemical basis, relevant lab tests, interpretation of results, and applied aspects of disorders of neonate and childhood.

Condition	Biochemical Basis	Relevant Lab Tests	Interpretation of Results
1. Congenital Thyroid Disease	Deficiency or absence of thyroid hormones due to congenital thyroid gland abnormalities.	TSH, Free T4, T3, Thyroid ultrasound, Genetic testing	Elevated TSH with low Free T4 indicates congenital hypothyroidism. Thyroid ultrasound helps in anatomical assessment.
2. Cystic Fibrosis	A genetic disorder caused by mutations in the CFTR gene, affecting chloride channels.	Sweat chloride test, Genetic testing, Immunoreactive Trypsinogen (IRT) Test	Elevated sweat chloride levels confirm cystic fibrosis. Genetic testing identifies CFTR mutations.
3. Delayed Puberty	Hormonal imbalances, including gonadotropin deficiency or resistance.	LH, FSH, Estradiol (females), Testosterone (males), and Bone age assessment	Low LH/FSH levels suggest hypogonadotropic hypogonadism. Elevated LH/FSH with low sex steroids indicates gonadal failure.

(Table 2) cont.....

Condition	Biochemical Basis	Relevant Lab Tests	Interpretation of Results
4. Diabetes Mellitus Type 1	Autoimmune destruction of pancreatic beta cells leading to insulin deficiency.	Fasting blood glucose, HbA1c, C-peptide, and Autoantibody tests (GAD, IA-2, ZnT8)	Elevated fasting glucose and HbA1c indicate diabetes. Low C-peptide suggests reduced insulin production.
5. Disorders of Sex Development (DSD), including Congenital Adrenal Hyperplasia (CAH)	Genetic mutations affecting sexual differentiation; CAH due to 21-hydroxylase deficiency.	17-Hydroxyprogesterone, Androstenedione, ACTH stimulation test, and Karyotyping	Elevated 17-hydroxyprogesterone indicates CAH. Karyotyping identifies chromosomal abnormalities.
6. Growth Retardation and Growth Hormone Deficiency	Lack of adequate growth hormone production or action.	Serum IGF-1 and Growth hormone stimulation tests (*e.g.*, insulin tolerance test)	Low IGF-1 and insufficient GH response to stimulation tests indicate GH deficiency.
7. Hypocalcemia	Low serum calcium levels due to hypoparathyroidism, vitamin D deficiency, or renal issues.	Serum calcium, Ionized calcium, PTH, Vitamin D levels, and Magnesium	Low calcium with low PTH suggests hypoparathyroidism. Low vitamin D levels indicate deficiency.
8. Inborn Errors of Metabolism	Genetic defects in metabolic pathways, leading to toxic metabolite accumulation.	Metabolic screening, Amino acid analysis, Organic acid analysis, and Enzyme assays	Abnormal metabolite levels indicate specific metabolic disorders. Genetic testing can confirm the diagnosis.
9. Neonatal Hypoglycemia	Low blood glucose levels due to inadequate glycogen stores or hormonal imbalances.	Blood glucose, Insulin levels, Cortisol, Growth hormone, and Critical sample analysis	Low blood glucose with high insulin indicates hyperinsulinism. Low cortisol or GH suggests a hormonal deficiency.
10. Neonatal Jaundice	Excess bilirubin due to immature liver function or hemolysis.	Total and direct bilirubin, Blood type, and Coombs test, G6PD deficiency screening	Elevated indirect bilirubin suggests physiological jaundice. Direct bilirubin elevation indicates possible cholestasis.
11. Neuroblastoma	Cancer arising from neural crest cells, often secreting catecholamines.	Urine Vanillylmandelic Acid (VMA), Homovanillic Acid (HVA), and Imaging studies	Elevated VMA/HVA levels suggest neuroblastoma. Imaging confirms tumor location and spread.

(Table 2) cont.....

Condition	Biochemical Basis	Relevant Lab Tests	Interpretation of Results
12. Precocious Puberty	Early activation of the hypothalamic-pituitary-gonadal axis or peripheral sources of sex steroids.	LH, FSH, Estradiol (females), Testosterone (males), and GnRH stimulation test	Elevated sex steroids with response to GnRH indicate central precocious puberty. Non-response suggests a peripheral cause.
13. Respiratory Distress	Difficulty in breathing due to various causes, including infections, prematurity, and structural abnormalities.	Blood gases, Chest X-ray, and Complete Blood Count (CBC)	Abnormal blood gases indicate impaired gas exchange. Imaging can reveal structural abnormalities.

LABORATORY INVESTIGATIONS

Relevant Laboratory Tests for Detection, Diagnosis, and Management, or Monitoring Analytics Methods Available and Applied Aspects for

Laboratory Investigations, Analytical Methods and Interpretation of Analytical Results (Table 3) [1 - 4]

Table 3. Laboratory investigations, analytical methods, and interpretation of disorders of neonate and childhood.

Test	Analytical Methods	Applied Aspects
Acid-Base/Blood Gases	Arterial blood gas analysis, pH measurement, pCO_2 and pO_2 measurement, and bicarbonate levels	Assessment of respiratory and metabolic function, critical in managing respiratory distress, metabolic acidosis/alkalosis, and assessing the adequacy of ventilation.
Calcium	Colorimetric assay, Ion-selective electrodes, and Atomic absorption spectroscopy	Monitoring bone health, parathyroid function, calcium metabolism disorders, and hypocalcemia, which can affect neuromuscular activity.
Glucose	Glucose oxidase method, Hexokinase method, and Point-of-care glucose meters	Essential for diagnosing and monitoring diabetes mellitus, neonatal hypoglycemia, and other metabolic disorders involving glucose metabolism.
Hormone Measurement		
-Anterior Hypothalamic Hormones	Immunoassays (*e.g.*, ELISA, RIA) and Mass spectrometry	Evaluating pituitary function, diagnosing hormone deficiencies (*e.g.*, growth hormone deficiency), and disorders like precocious and delayed puberty.

(Table 3) cont.....

Test	Analytical Methods	Applied Aspects
-Cortisol	Immunoassays and Liquid Chromatography-Mass Spectrometry (LC-MS/MS)	Assessing adrenal function, diagnosing adrenal insufficiency or hypercortisolism (*e.g.*, Cushing's syndrome).
-Neuropeptides	ELISA, Radioimmunoassay (RIA), and LC-MS/MS	Used for research purposes and in clinical trials to study their role in neuroendocrine functions, stress response, and metabolic regulation.
-Ovarian Hormones	Immunoassays and LC-MS/MS	Assessment of ovarian function, monitoring fertility, diagnosing disorders like Polycystic Ovarian Syndrome (PCOS), and managing reproductive health issues.
-Testicular Hormones	Immunoassays and LC-MS/MS	Evaluating testicular function, diagnosing hypogonadism, and assessing disorders of sex development.
-Thyroid Function	Immunoassays for TSH, Free T4, and Free T3	Diagnosing congenital thyroid disease, hypothyroidism, and hyperthyroidism, and monitoring treatment effectiveness.
-Other Adrenal Hormones	Immunoassays and LC-MS/MS	Assessment of adrenal cortex function, diagnosis of congenital adrenal hyperplasia, and other adrenal disorders.
HVA, Dopamine	High-Performance Liquid Chromatography (HPLC) and mass spectrometry	Diagnosing and monitoring neuroblastoma, as well as other disorders affecting catecholamine metabolism.
Performance of the Sweat Test	Pilocarpine iontophoresis and Sweat chloride measurement	Diagnosing cystic fibrosis and evaluating sweat electrolyte composition, which is critical for early detection and management of the disease.
Plasma Catecholamines	HPLC and LC-MS/MS	Diagnosis and management of pheochromocytoma, neuroblastoma, and other disorders affecting sympathetic nervous system activity.
Total and Direct Bilirubin	Colorimetric assay, Diazo method, and HPLC	Diagnosing and managing neonatal jaundice, liver function disorders, and hemolytic diseases.

SUMMARY

Pediatric clinical chemistry differs from adult clinical chemistry due to dynamic physiological changes in children, leading to varying reference ranges for biochemical markers. Laboratory testing often involves smaller sample volumes and different collection techniques. Accurate diagnosis and management require understanding these nuances and applying appropriate diagnostic tests. Conditions such as neonatal jaundice, inborn errors of metabolism, and electrolyte imbalances illustrate the complexity of pediatric clinical chemistry.

EXERCISES

- Define the term "pediatric reference range" and explain why these ranges are crucial for pediatric clinical chemistry.
- What are the main challenges associated with collecting and analyzing blood samples in neonates and infants?
- Describe the biochemical basis and laboratory diagnosis of neonatal jaundice.
- Explain the role of sweat chloride testing in the diagnosis of cystic fibrosis.
- Discuss the importance of monitoring blood glucose levels in pediatric patients with type 1 diabetes.
- What factors must be considered when interpreting thyroid function tests in neonates?
- What are the common causes and biochemical implications of hypocalcemia in neonates?
- Describe the biochemical basis for respiratory distress in newborns and the relevant laboratory investigations.
- What are the key biochemical markers used in the diagnosis of Inborn Errors of Metabolism (IEM), and how are they interpreted?
- Discuss the significance of neonatal hypoglycemia, its causes, and the biochemical markers involved.

CONCLUSION

To ensure effective pediatric clinical chemistry, a tailored approach must be implemented that takes into account the unique physiological characteristics of children. It is imperative to have a thorough understanding of proper sample collection, pediatric reference ranges, and the skills to interpret laboratory results. Continuous education and training are non-negotiable to stay abreast of advances in diagnostic technology and to ensure accurate and timely diagnosis, which is essential for optimal pediatric patient care.

REFERENCES

[1] Strasinger, S.K.; Di Lorenzo, M.S. *Clinical Chemistry: A Laboratory Perspective*; F.A. Davis Company, **2021**.

[2] Tahmasebi, H.; Higgins, V.; Fung, A.W.S.; Truong, D.; White-Al Habeeb, N.M.A.; Adeli, K. Pediatric Reference Intervals for Biochemical Markers: Gaps and Challenges, Recent National Initiatives and Future Perspectives. *EJIFCC,* **2017**, *28*(1), 43-63.
[PMID: 28439218]

[3] Kaplan, L.A.; Pesce, A.J.; Kazmierczak, S.C. *Clinical Chemistry: Theory, Analysis, Correlation*; 9th edition; **2026**.

[4] Petry, C.J. Reference intervals for laboratory tests and procedures in children. *Pediatr. Clin. North Am.,* **2020**, *67*(4), 681-694.

CHAPTER 19

Pregnancy and Prenatal Diagnosis

Abstract: Pregnancy and prenatal diagnosis encompass a range of biochemical and molecular changes that are critical for maternal and fetal health. This chapter reviews the biochemical markers and diagnostic techniques utilized in monitoring pregnancy and diagnosing prenatal conditions. It covers the role of hormones, proteins, and other biomarkers in assessing the health of the mother and fetus. Additionally, the chapter explores various prenatal diagnostic techniques, including Non-Invasive Prenatal Testing (NIPT), ultrasound, and amniocentesis, highlighting their importance in detecting genetic and chromosomal abnormalities. The integration of advanced molecular diagnostics and clinical chemistry is crucial for effective prenatal care and improved pregnancy outcomes.

LEARNING OBJECTIVES

After reading this chapter, one should be able to:

- Describe the following concepts about laboratory medicine support during pregnancy and related to the health of the mother and fetus.
- Describe the causes, clinical signs, and symptoms of the complications of pregnancy and fetal development.
- Describe the laboratory investigations that are important in their detection, diagnosis, and management.
- Describe specific laboratory investigations important to the study of pregnancy and fetal development.
- Recognize the analytical methods available for their measurement.
- Select laboratory investigations and interpret the analytical results in the context of clinical signs and symptoms.
- Discuss the clinical rationale for tests using amniotic fluid.

Simmi Kharb

KEY CONCEPTS

Key Concepts and Applied Aspects of Laboratory Medicine Support Related to Maternal and Fetal Health During Pregnancy

Laboratory Medicine Support Related to the Health of the Mother and Fetus During Pregnancy (Table 1)

Table 1. Laboratory medicine support related to the health of the mother and fetus during pregnancy.

Key Concepts	Description	Applied Aspects
Biochemical Changes	Pregnancy induces hormonal and metabolic changes, including increases in human Chorionic Gonadotropin (hCG), estrogen, and progesterone.	Monitoring hormone levels (*e.g.*, hCG, progesterone) for pregnancy confirmation, ectopic pregnancy, and miscarriage risk.
Hematological Changes	Pregnancy leads to physiological anemia due to increased plasma volume and hemodilution.	Complete Blood Count (CBC) to monitor hemoglobin levels, red cell mass, and identify anemia; iron supplementation if necessary.
Endocrine Changes	Endocrine adaptations include altered thyroid function and increased insulin resistance.	Thyroid function tests (T3, T4, TSH) to monitor maternal thyroid status; glucose tolerance tests to screen for gestational diabetes.
Prenatal Screening and Diagnosis	Involves biochemical markers (*e.g.*, PAPP-A, free β-hCG), ultrasound, and genetic testing for assessing fetal abnormalities.	First-trimester screening, Non-Invasive Prenatal Testing (NIPT) for Down syndrome, trisomy 18, and trisomy 13; amniocentesis and Chorionic Villus Sampling (CVS) for diagnosis.
Monitoring Maternal Health	Regular monitoring to detect and manage conditions like gestational diabetes, preeclampsia, and infections.	Blood pressure measurement, urine protein testing, glucose screening, and monitoring of infection markers (*e.g.*, TORCH panel).
Monitoring Fetal Health	Assessment of fetal growth, development, and well-being through biochemical and imaging methods.	Ultrasound for fetal growth and anatomy; Doppler studies for placental blood flow; fetal heart rate monitoring; amniotic fluid analysis.
Immunological Changes	Pregnancy involves modulation of the maternal immune system to prevent fetal rejection and manage immune responses.	Monitoring of immune markers (*e.g.*, IL-10, TNF-α) to understand immune tolerance mechanisms; RhD antibody testing in Rh-negative mothers.
Nutritional Status	Nutritional demands increase during pregnancy, requiring adequate intake of vitamins and minerals.	Monitoring of vitamin D, calcium, folate, and iron levels; supplementation as required to prevent deficiencies and ensure fetal development.

(Table 1) cont.....

Key Concepts	Description	Applied Aspects
Infectious Disease Screening	Screening for infections that can affect pregnancy outcomes, such as rubella, HIV, syphilis, and hepatitis.	Routine testing for Sexually Transmitted Infections (STIs) and other pathogens during early pregnancy to prevent transmission to the fetus.
Postpartum Monitoring	Assessment of maternal recovery and infant health following delivery.	Monitoring for postpartum anemia, thyroid function changes, and lactation issues; newborn screening for metabolic and genetic disorders.

Laboratory Tests During Pregnancy

Laboratory Investigations for Complications of Pregnancy and Fetal Development (Table 2 - 6)

Table 2. Laboratory investigations for complications of pregnancy and fetal development.

Test	Purpose	Analytical Methods	Interpretation
Complete Blood Count (CBC)	Monitor anemia, infection, and clotting status	Automated hematology analyzers	Low Hb: anemia; High WBC: infection; Platelets: clotting status
Glucose Tolerance Test (GTT)	Screen for gestational diabetes	Oral Glucose Tolerance Test (OGTT)	High glucose levels indicate gestational diabetes
Thyroid Function Tests (TSH, Free T4)	Monitor thyroid status	Immunoassays	High TSH: hypothyroidism; Low TSH: hyperthyroidism
Urinalysis	Screen for urinary tract infections, proteinuria	Dipstick analysis and microscopy	Proteinuria: preeclampsia; Bacteriuria: UTI
Liver Function Tests (ALT, AST, ALP)	Assess liver function	Enzymatic assays	Elevated levels may indicate liver dysfunction
Serum Electrolytes	Monitor electrolyte balance	Ion-Selective Electrodes (ISE)	Imbalances may indicate hyperemesis, preeclampsia
Hepatitis B Surface Antigen (HBsAg)	Screen for hepatitis B infection	Enzyme-linked immunosorbent assay	Positive: Hepatitis B infection
Rubella IgG	Assess immunity to rubella	Immunoassays	Positive: immunity; Negative: susceptible
HIV Antibody	Screen for HIV infection	ELISA and Western blot	Positive: HIV infection
Group B Streptococcus (GBS) Culture	Screen for GBS infection in late pregnancy	Culture methods	Positive: GBS colonization

Table 3. Biochemical changes during pregnancy.

Aspect	Details
Key Concepts	-**Metabolic Changes**: Increased basal metabolic rate, altered glucose metabolism, and lipid changes.
Relevant Tests	-**Glucose Tolerance Test (GTT)**: Screening for gestational diabetes -- **Lipid Profile**: Monitoring changes in cholesterol and triglycerides -- **Electrolytes**: Assessing sodium, potassium, calcium, and magnesium levels.
Interpretation	-**GTT**: Elevated glucose levels indicate gestational diabetes -- **Lipid Profile**: Increased triglycerides and cholesterol levels are normal in pregnancy -- **Electrolytes**: Slight fluctuations are common, but significant deviations may indicate underlying issues.
Applied Aspects	-**Management**: Monitor and manage gestational diabetes and dyslipidemia -**Nutritional Support**: Ensure adequate intake of essential minerals and vitamins.

Table 4. Hematological changes during pregnancy.

Aspect	Details
Key Concepts	-**Blood Volume**: Increase in blood volume by 30-50% -- **Red Blood Cells**: Increased RBC mass but also increased plasma volume, leading to physiological anemia -**White Blood Cells**: Increase in WBC count, particularly neutrophils -- **Coagulation**: Hypercoagulable state due to increased clotting factors.
Relevant Tests	-**Complete Blood Count (CBC)**: Monitoring hemoglobin, hematocrit, WBC, and platelet count -- **Coagulation Profile**: PT, aPTT, and fibrinogen levels -- **Iron Studies**: Ferritin, serum iron, and Total Iron-Binding Capacity (TIBC).
Interpretation	-**CBC**: Mild anemia is normal; significant anemia requires investigation -**Coagulation Profile**: Elevated fibrinogen and other clotting factors; careful monitoring in patients with thrombophilia -- **Iron Studies**: Low ferritin indicates iron deficiency; TIBC may be increased.
Applied Aspects	-**Management**: Iron supplementation for anemia, anticoagulation therapy in high-risk patients -- **Monitoring**: Regular CBC and coagulation profile checks, especially in patients with a history of bleeding or clotting disorders.

Table 5. Endocrine changes during pregnancy.

Aspect	Details
Key Concepts	-**Hormonal Changes**: Increased levels of human Chorionic Gonadotropin (hCG), progesterone, and estrogen -- **Thyroid Function**: Increased thyroid hormone production, increased total T3 and T4 levels due to increased Thyroid-Binding Globulin (TBG), but free T3 and T4 levels remain unchanged or slightly increased -- **Glucose Metabolism**: Increased insulin resistance, particularly in the second and third trimesters.

(Table 5) cont.....

Aspect	Details
Relevant Tests	-**hCG Levels**: Monitoring for normal pregnancy progression -- **Thyroid Function Tests**: TSH, free T3, free T4, and thyroid antibodies- -**Oral Glucose Tolerance Test (OGTT)**: Screening for gestational diabetes.
Interpretation	-**hCG Levels**: Rapid increase in early pregnancy; abnormally low or high levels may indicate pregnancy complications -- **Thyroid Function Tests**: Slightly lower TSH in the first trimester; significant changes may indicate thyroid dysfunction -- **OGTT**: Elevated glucose levels after the test indicate gestational diabetes.
Applied Aspects	-**Management**: Thyroid hormone replacement for hypothyroidism, anti-thyroid medications for hyperthyroidism -- **Monitoring**: Regular thyroid function tests and glucose monitoring to manage gestational diabetes and thyroid disorders.

Table 6. Maternal and fetal health.

Aspect	Details
Key Concepts	-**Prenatal Screening**: Testing for genetic and congenital abnormalities -- **Monitoring Fetal Well-being**: Ultrasound, fetal heart rate monitoring -- **Maternal Health**: Monitoring for complications such as preeclampsia, gestational diabetes, and infections.
Relevant Tests	-**First Trimester Screening**: hCG, Pregnancy-Associated Plasma Protein-A (PAPP-A), nuchal translucency ultrasound- -**Second Trimester Screening**: Quad screen (AFP, hCG, estriol, inhibin-A) -- **Non-Invasive Prenatal Testing (NIPT)**: Cell-free fetal DNA testing for chromosomal abnormalities -- **Invasive Testing**: Amniocentesis and Chorionic Villus Sampling (CVS) for genetic testing.
Interpretation	-**Screening Tests**: Abnormal results warrant further diagnostic testing -**NIPT**: High accuracy for detecting common chromosomal abnormalities like Down syndrome -- **Invasive Testing**: Confirmatory diagnosis of genetic abnormalities.
Applied Aspects	-**Management**: Counseling and management of identified genetic conditions, monitoring and managing maternal health conditions, and timely intervention for fetal distress.

DISEASE

Biochemical Basis, Relevant Lab Tests, Interpretation, and Applied Aspects

Laboratory Investigations for Complications of Pregnancy and Fetal Development, Interpretation and Applied Aspects (Tables 7-16) [1 - 5]

Table 7. Biochemical basis, relevant lab tests, interpretation, and applied aspects: Choriocarcinoma.

Aspect	Details
Biochemical Basis	Malignant trophoblastic tumor arising from placental tissue, characterized by high hCG levels.
Relevant Lab Tests	-**Serum hCG Levels**: Quantitative hCG testing.
Interpretation	-**Elevated hCG**: Extremely high levels are indicative of choriocarcinoma.
Applied Aspects	-**Management**: Chemotherapy, monitoring hCG levels post-treatment for recurrence.

Table 8. Biochemical basis, relevant lab tests, interpretation, and applied aspects: Ectopic Pregnancy.

Aspect	Details
Biochemical Basis	Implantation of the fertilized egg outside the uterine cavity, often in the fallopian tube.
Relevant Lab Tests	-**Serum hCG Levels**: Quantitative hCG testing -- **Progesterone Levels**.
Interpretation	-**hCG**: Lower-than-normal rise in hCG levels -- **Progesterone**: Low levels.
Applied Aspects	-**Management**: Surgical intervention or methotrexate; monitoring for hCG decline.

Table 9. Biochemical basis, relevant lab tests, interpretation, and applied aspects: Gestational Diabetes.

Aspect	Details
Biochemical Basis	Increased insulin resistance during pregnancy, leading to hyperglycemia.
Relevant Lab Tests	-**Oral Glucose Tolerance Test (OGTT)** -- **Fasting Blood Glucose** -- **HbA1c**.
Interpretation	-**OGTT**: Elevated glucose levels post-ingestion -- **Fasting Glucose**: High fasting levels -- **HbA1c**: Elevated.
Applied Aspects	-**Management**: Diet, exercise, insulin therapy if needed; monitoring glucose levels.

Table 10. Biochemical basis, relevant lab tests, interpretation, and applied aspects: Molar Pregnancy.

Aspect	Details
Biochemical Basis	Abnormal growth of trophoblastic tissue; partial or complete molar pregnancies.
Relevant Lab Tests	-**Serum hCG Levels**: Quantitative hCG testing.
Interpretation	-**Elevated hCG**: Extremely high levels suggest molar pregnancy.
Applied Aspects	-**Management**: Surgical removal of molar tissue; monitoring hCG levels post-removal.

Table 11. Biochemical basis, relevant lab tests, interpretation, and applied aspects: Obstetric Cholestasis.

Aspect	Details
Biochemical Basis	Liver disorder during pregnancy leading to elevated bile acids and liver enzymes.
Relevant Lab Tests	-**Serum Bile Acids** -- **Liver Function Tests (LFTs)**: ALT, AST, and bilirubin.
Interpretation	-**Bile Acids**: Elevated levels -- **LFTs**: Elevated liver enzymes and bilirubin.
Applied Aspects	-**Management**: Ursodeoxycholic acid, early delivery if severe; monitoring LFTs and bile acids.

Table 12. Biochemical basis, relevant lab tests, interpretation, and applied aspects: Open neural tube defects.

Aspect	Details
Biochemical Basis	Failure of neural tube closure during embryonic development, leading to defects like spina bifida.
Relevant Lab Tests	-**Maternal Serum Alpha-Fetoprotein (MSAFP)** -- **Ultrasound**.
Interpretation	-**MSAFP**: Elevated levels suggest neural tube defects -- **Ultrasound**: Visual confirmation.
Applied Aspects	-**Management**: Prenatal surgery or planning for postnatal care; folic acid supplementation.

Table 13. Biochemical basis, relevant lab tests, interpretation, and applied aspects: Pre-Eclampsia, HELLP syndrome.

Aspect	Details
Biochemical Basis	Pre-eclampsia: High blood pressure and proteinuria during pregnancy-HELLP: Hemolysis, Elevated Liver enzymes, Low Platelets.

(Table 13) cont.....

Aspect	Details
Relevant Lab Tests	-**Blood Pressure Monitoring** -- **Urinalysis**: Proteinuria -- **LFTs** -- **Platelet Count** -- **LDH**.
Interpretation	-**BP**: Elevated -- **Urinalysis**: Proteinuria -- **LFTs**: Elevated liver enzymes -- **Platelets**: Low -- **LDH**: Elevated.
Applied Aspects	-**Management**: Antihypertensives, magnesium sulfate, delivery planning; close monitoring of mother and fetus.

Table 14. Biochemical basis, relevant lab tests, interpretation, and applied aspects: Rh isoimmunization.

Aspect	Details
Biochemical Basis	Maternal immune response against fetal Rh-positive blood cells, leading to hemolytic disease of the newborn.
Relevant Lab Tests	-**Blood Typing and Rh Factor** -- **Indirect Coombs Test**: Detect maternal antibodies.
Interpretation	-**Indirect Coombs**: Positive indicates the presence of antibodies.
Applied Aspects	-**Management**: Anti-D immunoglobulin (RhoGAM) administration; monitoring for fetal anemia and intervention if necessary.

Table 15. Biochemical basis, relevant lab tests, interpretation, and applied aspects: Trisomy 21, 18, and 13.

Aspect	Details
Biochemical Basis	Genetic disorders resulting from extra copies of chromosomes 21 (Down syndrome), 18 (Edwards syndrome), or 13 (Patau syndrome).
Relevant Lab Tests	-**First Trimester Screening**: hCG, PAPP-A -- **NIPT**: Cell-free fetal DNA -- **Amniocentesis** or **CVS**.
Interpretation	-**Screening**: Abnormal results indicate a higher risk -- **NIPT**: High accuracy for detecting trisomies -- **Amniocentesis/CVS**: Confirmatory diagnosis.
Applied Aspects	-**Management**: Genetic counseling, prenatal and postnatal care planning.

Table 16. Biochemical basis, relevant lab tests, interpretation, and applied aspects: Trophoblastic disease.

Aspect	Details
Biochemical Basis	Spectrum of disorders arising from abnormal placental trophoblast cells, including molar pregnancy and choriocarcinoma.
Relevant Lab Tests	-**Serum hCG Levels**: Quantitative hCG testing.
Interpretation	-**Elevated hCG**: Extremely high levels suggest trophoblastic disease.
Applied Aspects	-**Management**: Surgical intervention, chemotherapy for malignant forms, and regular hCG monitoring post-treatment.

DISORDERS

Biochemical Basis, Relevant Lab Tests, Interpretation, and Applied Aspects

Biochemical Basis, Relevant Lab Tests, Interpretation, and Applied Aspects (Tables 17 - 27) [1 - 5]

Table 17. Changes in analyte concentrations throughout pregnancy: Multiples of the Median (MoM).

Aspect	Details
Biochemical Basis	Changes in levels of various analytes as pregnancy progresses, influenced by placental and fetal development.
Relevant Lab Tests	**-hCG** **-AFP** **-Estriol** **-Inhibin-A** **-PAPP-A**
Interpretation	-**MoM**: Used to compare individual test results to the median expected at a given gestational age.
Applied Aspects	-**Monitoring**: Regular screening for abnormal levels, which may indicate potential complications.

Table 18. Biochemical basis, relevant lab tests, interpretation, and applied aspects: *Fetal lung maturity*.

Aspect	Details
Biochemical Basis	Measurement of surfactant production, crucial for reducing surface tension in the lungs.
Relevant Lab Tests	**-Lecithin/Sphingomyelin (L/S) Ratio-Phosphatidylglycerol (PG)-Foam Stability Index**
Interpretation	-**L/S Ratio > 2:1**: Indicates maturity -- **Presence of PG**: Indicates maturity.

(Table 18) cont.....

Aspect	Details
Applied Aspects	-**Management**: Decision-making for delivery timing in cases of preterm labor.

Table 19. Biochemical basis, relevant lab tests, interpretation, and applied aspects: *hCG doubling time.*

Aspect	Details
Biochemical Basis	Measurement of the time it takes for hCG levels to double, used to assess the viability of the pregnancy.
Relevant Lab Tests	**-Serial hCG Measurements**
Interpretation	-**Doubling Time of 48-72 Hours**: Normal in early pregnancy -- **Slower/Faster Doubling**: May indicate abnormality.
Applied Aspects	-**Monitoring**: Tracking hCG levels to assess pregnancy viability.

Table 20. Biochemical basis, relevant lab tests, interpretation, and applied aspects: *hCG Forms/Variants.*

Aspect	Details
Biochemical Basis	Different molecular forms of hCG produced by the placenta, including intact hCG, free beta subunit, and hyperglycosylated hCG.
Relevant Lab Tests	**-Total hCG** **-Free Beta hCG**
Interpretation	-**Specific Forms**: Different forms may be used to monitor different aspects of pregnancy and trophoblastic diseases.
Applied Aspects	-**Differential Diagnosis**: Identifying and managing conditions like molar pregnancy and choriocarcinoma.

Table 21. Biochemical basis, relevant lab tests, interpretation, and applied aspects: Maternal serum screening: First trimester screening.

Aspect	Details
Purpose	Early detection of chromosomal abnormalities like Down syndrome.
Relevant Lab Tests	**-hCG** **-PAPP-A** **-Nuchal Translucency** **- Ultrasound**
Interpretation	-**High hCG and Low PAPP-A**: Increased risk for Down syndrome.
Limitations	-**Screening Test**: Indicates risk, not definitive -- **Follow-up**: Requires diagnostic testing if positive.
Applied Aspects	-**Counseling**: Explaining results and potential next steps to expectant parents.

Table 22. Biochemical basis, relevant lab tests, interpretation, and applied aspects: Maternal serum screening: Integrated screening.

Aspect	Details
Purpose	Combines first- and second-trimester tests for better accuracy.
Relevant Lab Tests	-**First Trimester**: hCG, PAPP-A, and nuchal translucency -- **Second Trimester**: AFP, estriol, hCG, and inhibin-A.
Interpretation	-**Combined Risk Assessment**: More accurate risk prediction for chromosomal abnormalities.
Limitations	-**Timing**: Requires completion of both test stages for the final result.
Applied Aspects	-**Follow-up**: Diagnostic tests if integrated screening indicates high risk.

Table 23. Biochemical basis, relevant lab tests, interpretation, and applied aspects: Maternal serum screening: Second trimester screening.

Aspect	Details
Purpose	Screening for chromosomal abnormalities and neural tube defects.
Relevant Lab Tests	-**Quad Screen**: AFP, hCG, estriol, and inhibin-A
Interpretation	-**High AFP**: Risk for neural tube defects -- **Abnormal Quad**: Risk for Down syndrome and trisomy 18.
Limitations	-**Screening**: Indicates risk, requires follow-up diagnostic testing if positive.
Applied Aspects	-**Counseling and Diagnostic Testing**: Amniocentesis or detailed ultrasound if screening is abnormal.

Table 24. Biochemical basis, relevant lab tests, interpretation, and applied aspects: Maternal serum screening: Non-Invasive Prenatal Testing (NIPT)/cell-free DNA testing.

Aspect	Details
Purpose	Early detection of chromosomal abnormalities using cell-free fetal DNA in maternal blood.
Relevant Lab Tests	-**cfDNA Testing**
Interpretation	-**High Accuracy**: For detecting trisomy 21, 18, 13.
Limitations	-**False Positives/Negatives**: Rare but possible, requires confirmatory testing.
Applied Aspects	-**Follow-up**: Diagnostic testing, like amniocentesis, if NIPT indicates high risk.

Table 25. Biochemical basis, relevant lab tests, interpretation, and applied aspects: Maternal serum screening: Premature Rupture of Membranes (PROM) and pre-term labor.

Aspect	Details
Biochemical Basis	Rupture of the amniotic sac before labor begins, increasing the risk of infection and preterm birth.
Relevant Lab Tests	-**AmniSure Test**: Detects Placental Alpha Microglobulin-1 (PAMG-1) -- **Nitrazine Test**: pH testing of vaginal fluid -- **Fern Test**: Microscopic examination for ferning pattern.
Interpretation	-**Positive AmniSure/Nitrazine/Fern Test**: Indicates rupture of membranes.
Applied Aspects	-**Management**: Hospitalization, antibiotics, corticosteroids to enhance fetal lung maturity, and monitoring for signs of labor.

Table 26. Biochemical basis, relevant lab tests, interpretation, and applied aspects: Maternal serum screening: Preterm birth prediction.

Aspect	Details
Biochemical Basis	Identifying biomarkers and risk factors for early labor onset.
Relevant Lab Tests	-**Fetal Fibronectin (fFN)**: Cervicovaginal secretion test -- **Cervical Length Measurement**: Transvaginal ultrasound.
Interpretation	-**Positive fFN**: Increased risk of preterm birth -- **Short Cervical Length**: Increased risk of preterm birth.
Applied Aspects	-**Management**: Interventions to prevent preterm birth, such as progesterone supplementation, cervical cerclage, and lifestyle modifications.

Table 27. Biochemical basis, relevant lab tests, interpretation, and applied aspects: Maternal Serum Screening: Rh isoimmunization.

Aspect	Details
Biochemical Basis	Maternal immune response against fetal Rh-positive blood cells, leading to hemolytic disease of the newborn.
Relevant Lab Tests	-**Blood Typing and Rh Factor-Indirect Coombs Test**: Detect maternal antibodies.
Interpretation	-**Positive Indirect Coombs**: Indicates presence of Rh antibodies.
Applied Aspects	-**Management**: Administer anti-D immunoglobulin (RhoGAM) to prevent sensitization; monitor for fetal anemia and consider intrauterine transfusions if necessary.

LABORATORY INVESTIGATIONS

Relevant Laboratory Tests for the Detection, Diagnosis, Management, or Monitoring During Pregnancy

Laboratory Tests for the Detection, Diagnosis, Management, or Monitoring During Pregnancy (Table 28)

Table 28. Laboratory tests for the detection, diagnosis, management, or monitoring during pregnancy.

Laboratory Test	Purpose	Tests	Applied Aspects
AFP (Alpha-Fetoprotein)	Screening for neural tube defects and chromosomal abnormalities.	- Maternal Serum AFP -Amniotic Fluid AFP	- Routine screening - Diagnostic confirmation for elevated levels.
hCG (Human Chorionic Gonadotropin)	Early pregnancy detection and monitoring of pregnancy health.	- Serum or Urine hCG	- Pregnancy confirmation - Monitoring pregnancy progression and health.
Inhibin A	Screening for Down syndrome and chromosomal abnormalities.	- Maternal Serum Inhibin A	- Combined with other markers for risk assessment.
PAPP-A (Pregnancy-Associated Plasma Protein A)	Early detection of Down syndrome and chromosomal abnormalities.	- Maternal Serum PAPP-A	- Combined with other markers for risk assessment.
Plasma Nucleic Acids (cfDNA Testing)	Non-Invasive Prenatal Testing (NIPT) for fetal chromosomal abnormalities.	- Cell-Free DNA in Maternal Blood	- High accuracy in detecting trisomies and sex chromosome aneuploidies.
Serum Bile Acids	Diagnosis and monitoring of obstetric cholestasis.	- Serum Bile Acids	- Diagnosis and management of obstetric cholestasis.
Unconjugated Estriol	Assessment of fetal well-being and placental function.	- Maternal Serum Unconjugated Estriol	- Part of the second-trimester maternal serum screening.

(Table 28) cont.....

Laboratory Test	Purpose	Tests	Applied Aspects
Amniotic Fluid Testing	Diagnostic tests performed on amniotic fluid.	- Acetylcholinesterase-AFP-Karyotype	- Confirmation of neural tube defects or chromosomal abnormalities identified through screening tests.

SUMMARY

Pregnancy induces significant physiological and biochemical changes in a woman's body, which are necessary for supporting fetal development and maintaining maternal health. Key biochemical markers, such as human Chorionic Gonadotropin (hCG), Alpha-Fetoprotein (AFP), and Pregnancy-Associated Plasma Protein-A (PAPP-A), are essential for monitoring pregnancy progression. Prenatal diagnosis, including techniques like NIPT, amniocentesis, and Chorionic Villus Sampling (CVS), allows for early detection of genetic and chromosomal abnormalities, facilitating informed decision-making and management of high-risk pregnancies. Understanding these biomarkers and diagnostic methods is vital for clinicians to ensure the health and well-being of both mother and child.

EXERCISES

- Which biochemical marker is commonly used to confirm pregnancy in its early stages?
- Non-Invasive Prenatal Testing (NIPT) is primarily used to detect which condition?
- What is the primary advantage of using Non-Invasive Prenatal Testing (NIPT) over traditional methods like amniocentesis?
- Describe the use of Maternal Serum Alpha-Fetoprotein (MSAFP) testing during pregnancy.
- Explain the significance of cell-free fetal DNA (cffDNA) in prenatal diagnosis and its impact on current screening practices.

CONCLUSION

The application of clinical chemistry in pregnancy and prenatal diagnosis is invaluable for monitoring and ensuring maternal and fetal health. The identification of specific biomarkers and the use of advanced diagnostic techniques enable early detection and management of potential complications. As technology advances, the integration of non-invasive and more accurate diagnostic methods will continue to improve prenatal care, offering safer and more effective options for patients. Ongoing research and ethical considerations will guide the future development of prenatal diagnostic practices.

REFERENCES

[1] Poulton, A.; Hui, L. Noninvasive prenatal testing: an overview. *Aust Prescr,* **2025**, *48*(2), 47-53. [http://dx.doi.org/10.18773/austprescr.2025.019]

[2] Audibert, F.; Wou, K.; Okun, N.; Bie, D.; Wilson, I.; Guideline, R.D. Prenatal Screening for Fetal Chromosomal Anomalies *J Obstet Gynaecol Can,* **2024**, *46*, 102694. [http://dx.doi.org/10.1016/j.jogc.2024.102694]

[3] Carbone, L; Cariati, F; Sarno, L; Conforti, A; Bagnulo, F; Strina, I **2020**. [http://dx.doi.org/10.3390/genes12010015]

[4] Swanson, K.; Norton, M.E. Expanded applications of cell-free fetal DNA screening. *Best Pract Res Clin Obstet Gynaecol,* **2025**, *98*, 102574. [http://dx.doi.org/10.1016/j.bpobgyn.2024.102574]

[5] Abedalthagafi, M. Non-invasive prenatal testing: a revolutionary journey in prenatal testing. *Front. Med. Sec. Obstetrics and Gynecology,* **2023**, 10. [http://dx.doi.org/10.3389/fmed.2023.1265090]

CHAPTER 20

Therapeutic Drug Monitoring and Toxicology

Abstract: Therapeutic Drug Monitoring (TDM) and toxicology are essential for personalized medicine, ensuring safe and effective medication use while minimizing toxicity. TDM involves measuring drug concentrations in the blood to optimize dosage, especially for drugs with narrow therapeutic windows. Toxicology focuses on identifying, managing, and preventing the toxic effects of drugs and chemicals. This chapter highlights the practical aspects of TDM and toxicology, emphasizing their importance in clinical decision-making and addressing future challenges and directions in these fields.

Keywords: Drug measurement, Pharmacodynamics, Pharmacokinetics, Therapeutic drug monitoring, Toxicology.

LEARNING OBJECTIVES

After going through this chapter, one should be able to:

- Describe key concepts of pharmacology and key factors relevant to drug action and measurement.
- Describe the mode of action and clinical uses of drugs in the categories listed as Pharmacokinetics (PK), Pharmacodynamics (PD), and Pharmacogenetics (PG).
- Describe the specific laboratory investigations and concepts important to therapeutic drug monitoring and toxicology.
- Recognize the analytical methods available for their measurement.
- Select laboratory investigations and interpret the analytical results in the context of clinical signs and symptoms.

Simmi Kharb

KEY CONCEPTS

Key Concepts of Pharmacology: Mode of Action and Clinical Uses of a Drug, Drug Action, and Measurement

Pharmacology and Key Factors Relevant to Drug Action and Measurement (Table 1)

Table 1. Drug action, its measurement, and applied aspects.

Key Concept	Details	Applied Aspects
Mode of Action of Drugs	Mechanism by which drugs exert their effects on the body at the molecular, cellular, or organ level.	- Understanding allows for targeted therapy - Basis for drug development and optimization.
Clinical Uses of Drugs	Therapeutic applications of drugs to treat diseases and conditions.	- Application in disease management and patient - Guiding treatment protocols and guidelines.
Drug Action	Biochemical and physiological effects of drugs on the body.	- Studying drug interactions and side effects - Optimizing dosage and administration regimens.
Measurement of Drug Action	Quantification of drug effects in biological systems.	- Pharmacokinetic and pharmacodynamic studies - Monitoring drug levels to ensure therapeutic efficacy.

Biochemical Basis, Relevant Lab Tests, Interpretation of Results, and Applied Aspects for

Laboratory Investigations, Analytical Methods and Interpretation of Analytical Results (Table 2) [1-3]

Table 2. Biochemical basis, relevant lab tests, analytical methods, and interpretation of results.

Category	Key Concepts	Applied Aspects
Therapeutic Drug Monitoring (TDM)	***Biochemical Basis***	
	Pharmacokinetics	Absorption, distribution, metabolism, and excretion of drugs
	Pharmacodynamics	Drug effects and mechanisms of action
	Therapeutic Range	Effective concentration range without toxicity
	Half-Life	Time required for the drug concentration to decrease by half
	Bioavailability	Proportion of the drug entering circulation to have an active effect
	Analytical Methods	
	High-Performance Liquid Chromatography (HPLC)	Separates and quantifies drug components
	Liquid Chromatography-Tandem Mass Spectrometry (LC-MS/MS)	Sensitive and specific for low drug concentrations
	Immunoassays	ELISA, RIA, fluorescence polarization immunoassay
	Gas Chromatography-Mass Spectrometry (GC-MS)	Used for volatile and non-volatile drugs
	Capillary Electrophoresis (CE)	Separation based on charge and size of molecules
	Interpretation of Results	
	Major Drug Classes	Anticonvulsants, antibiotics, immunosuppressants, anticoagulants, and cardiac drugs
	Rationale for TDM	Narrow therapeutic index, patient variability, drug interactions, compliance, and disease states
	Free *vs.* Total Drug Levels	Free drug is biologically active; relevant in altered protein binding
	Interpretive Guidelines Examples:	*Therapeutic ranges and cutoffs for specific drugs*
	Phenytoin	Therapeutic range: 10-20 μg/mL, Toxicity > 20 μg/mL
	Vancomycin	Trough level: 10-20 μg/mL, higher for severe infections
	Tacrolimus	Therapeutic range: 5-20 ng/mL depending on transplant type
	Warfarin (INR)	Target INR: 2.0-3.0

(Table 2) cont.....

Category	Key Concepts	Applied Aspects
Toxicology	***Biochemical Basis***	
	Toxin Metabolism	Biotransformation by the liver and other tissues
	Accumulation	Build-up of toxic substances in the body
	Biomarkers	Indicate exposure or effect of toxins
	Mechanisms of Toxicity	Cellular and molecular mechanisms causing harm
	Analytical Methods	
	Gas Chromatography-Mass Spectrometry (GC-MS)	To confirm the presence of drugs and toxins
	Liquid Chromatography-Mass Spectrometry (LC-MS/MS)	Used for non-volatile substances
	Atomic Absorption Spectroscopy (AAS)	To measure the concentrations of metals
	Inductively Coupled Plasma Mass Spectrometry (ICP-MS)	Detection of trace elements
	Immunoassays	Rapid screening for drugs of abuse and toxicants
	Enzyme Assays	Specific for certain toxins or metabolic byproducts
	Interpretation of Results	
	Investigation of Poisoning	Screening and confirmatory tests, clinical context guides testing
	Toxic Concentrations:	*Toxicity:*
	Phenytoin > 20 µg/mL	nystagmus, ataxia, seizures
	Vancomycin > 20 µg/mL	nephrotoxicity, ototoxicity
	Lithium > 1.5 mEq/L	toxicity: tremors, confusion, seizures
	Others:	
	Drug of Abuse Testing	Screening (immunoassays) and confirmatory (GC-MS or LC-MS/MS)
	Clinical Use	Monitoring compliance in rehabilitation, legal investigations, and emergency settings
Pharmacokinetics (PK)	***Factors***	

(Table 2) cont.....

Category	Key Concepts	Applied Aspects
	Absorption	The process by which a drug enters the bloodstream
	Bioavailability	Fraction of administered drug reaching systemic circulation
	Compliance	Adherence to prescribed medication regimens
	Distribution	Dispersion of drugs throughout the body
	Excretion	Removal of drugs from the body, primarily *via* the kidneys
	Metabolism	Biotransformation of drugs, mainly in the liver
	Steady State	Drug intake and elimination are balanced
	Immediate *vs* Extended Release	Differences in drug formulation, affecting absorption and duration of action
Pharmacodynamics (PD) and Pharmacogenetics (PG)	***Drug Classes***	
	Antibiotics	Monitoring efficacy and resistance
	Anticonvulsants/Antiepileptics	Ensuring therapeutic levels for seizure control
	Antidepressants	Monitoring levels to avoid toxicity and ensure efficacy
	Anti-infectives	Ensuring appropriate drug levels for infection control
	Anticoagulants	Monitoring INR for safe and effective anticoagulation
	Immunosuppressants/Chemotherapeutics	Ensuring therapeutic levels to prevent rejection and manage cancer treatment
Therapeutic Drug Monitoring and Toxicology	***Interpretation and Application***	

(Table 2) cont.....

Category	Key Concepts	Applied Aspects
	Therapeutic Drug Monitoring	Ensuring therapeutic efficacy and minimizing toxicity
	Toxicology Monitoring	Identifying and managing exposure to toxic substances
	Causes and Clinical Signs/Symptoms	
	Subtherapeutic Levels	Phenytoin (seizures), Warfarin (thromboembolism), Tacrolimus (organ rejection), and Vancomycin (infection)
	Toxic Levels	Phenytoin (nystagmus, ataxia, confusion, and seizures) Warfarin (bleeding), Tacrolimus (nephrotoxicity), and Vancomycin (nephrotoxicity)

Laboratory Tests, Analytics Methods Available, and Applied Aspects of Therapeutic Drug Monitoring (TDM)

Laboratory Investigations, Analytics Methods Available and Therapeutic Drug Monitoring and Toxicology (Table 3 - 5) [1 - 4]

Table 3. Laboratory tests, analytics methods, and interpretation and applied aspects of Therapeutic Drug Monitoring (TDM).

Aspect	*Details*	*Applied Aspects*
Biochemical Basis	Monitoring blood levels of drugs to ensure they are within the therapeutic range, avoiding subtherapeutic or toxic levels.	- Ensures drug efficacy and safety - Helps adjust dosage for individual patients.
Relevant Lab Tests	- Serum/plasma drug concentration measurements - High-Performance Liquid Chromatography (HPLC) - Immunoassays.	- Commonly monitored drugs: Anticonvulsants, antibiotics, immunosuppressants, and anticoagulants.
Interpretation of Results	-**Therapeutic Range**: Optimal drug concentration for efficacy with minimal side effects -**Subtherapeutic Levels**: Indicate inadequate dosing or poor compliance. -**Toxic Levels**: Indicate overdose or accumulation, requiring dosage adjustment or drug discontinuation.	- Guides dosage adjustments. - Identifies non-compliance or absorption issues - Prevents toxicity.

(Table 3) cont.....

Aspect	*Details*	*Applied Aspects*
Applied Aspects	-**Individualized Therapy**: Adjusts dosing based on individual patient metabolism and drug clearance -**Drug Interactions**: Identifies potential drug-drug interactions that affect drug levels -**Special Populations**: Adjusts dosing for pediatric, elderly, and renal/hepatic impaired patients.	- Optimizes drug therapy - Enhances patient safety - Improves therapeutic outcomes.

Table 4. TDM and toxicology applications.

Application	*Details*	*Examples*
Anticonvulsants (TDM)	Monitoring blood levels to maintain the therapeutic range and prevent seizures or toxicity.	- Phenytoin Valproate Carbamazepine
Antibiotics (TDM)	Ensuring effective drug levels to treat infections while avoiding toxicity.	- Vancomycin Aminoglycosides (*e.g.*, gentamicin, tobramycin)
Immunosuppressants (TDM)	Monitoring levels to prevent organ rejection while minimizing side effects.	- Cyclosporine Tacrolimus Sirolimus
Anticoagulants (TDM)	Monitoring to ensure appropriate anticoagulation and prevent bleeding or clotting.	- Warfarin (INR) Heparin (aPTT) Direct Oral Anticoagulants (DOACs)
Toxicology (Drugs of Abuse)	Detecting and quantifying drugs to manage overdoses and support legal cases.	- Cocaine Heroin Methamphetamine
Toxicology (Environmental Toxins)	Identifying exposure to harmful substances and guiding medical or occupational interventions.	- Lead Mercury Arsenic
Clinical Toxicology (Chronic Exposure)	Monitoring and managing patients with chronic exposure to toxins, ensuring long-term health and safety.	- Occupational exposure to solvents Chronic drug abuse

Table 5. Laboratory tests, analytics methods available, and applied aspects of toxicology.

Aspect	*Details*	*Applied Aspects*
Biochemical Basis	Study of the harmful effects of chemicals (including drugs) on living organisms and the mechanisms of these effects.	- Identifies and quantifies toxic substances in biological specimens. Determines the cause of toxicity.

(Table 5) cont.....

Aspect	*Details*	*Applied Aspects*
Relevant Lab Tests	-**Blood/Urine Tests**: Screening and confirmatory tests for drugs of abuse and toxic substances -**Chromatography**: HPLC, Gas Chromatography (GC), Mass Spectrometry (MS) -**Immunoassays**: Enzyme-Linked Immunosorbent Assay (ELISA), Radioimmunoassay (RIA).	- Identifies acute and chronic exposure to toxic substances - Monitors therapy for overdose or poisoning.
Interpretation of Results	-**Positive Results**: Indicates the presence of toxic substances or drugs -**Quantitative Levels**: Determines the extent of exposure or intoxication. -**Metabolite Analysis**: Identifies specific metabolites indicating drug use or poisoning.	- Guides treatment for overdose or poisoning - Legal and forensic implications - Occupational exposure monitoring.
Applied Aspects	-**Emergency Medicine**: Rapid identification of toxins for acute management. -**Forensic Toxicology**: Legal investigations related to drug use, poisoning, or exposure -**Clinical Toxicology**: Management of chronic exposure and substance abuse.	- Enhances patient care in emergency and chronic settings. - Supports legal investigations. - Improves public health and safety.

SUMMARY

The use of precise analytical methods ensures accurate measurement of drug levels and toxins, facilitating effective clinical management. Understanding the biochemical basis, interpretation of results, and clinical implications of TDM and toxicology testing is crucial for optimizing patient care and safety.

EXERCISES

- What is the primary goal of Therapeutic Drug Monitoring (TDM)?
- Which analytical methods are commonly used in TDM to optimize drug dosage for efficacy while minimizing toxicity?
- List three drugs that require regular therapeutic drug monitoring.
- Why is TDM particularly important for drugs with a narrow therapeutic index?
- What is the role of toxicology in emergency departments?
- Describe the principle behind High-Performance Liquid Chromatography (HPLC) in TDM.
- What is the significance of measuring plasma levels of digoxin in TDM?
- How do pharmacokinetic properties influence the decision to monitor drug levels?
- What are the main symptoms of lithium toxicity?
- Explain the role of antidotes in toxicology and provide an example.

CONCLUSION

Therapeutic drug monitoring and toxicology are crucial aspects of clinical practice that ensure the safe and effective use of medications, guide treatment in cases of overdose or poisoning, and support legal and occupational health investigations. These disciplines rely on accurate laboratory tests to provide critical information for patient management and public health.

REFERENCES

[1] Petrova, G.; Blagova, S.; Tachkov, K.; Santos, M.; Bluett, J.; Rumano, M.; Kkolou, E.; Drakalska, E.; Arev, M.; Barsbay, M.C.; Mulleman, D. Therapeutic drug monitoring education: The current state. *Br J Clin Pharmacol,* **2025**, *92*(1), 198-209. [http://dx.doi.org/10.1002/bcp.70252]

[2] Udomkarnjananun, S. Therapeutic drug monitoring of immunosuppressive drugs in hepatology and gastroenterology *Best Pract Res Clin Gastroenterol,* **2021**, 54-55. [http://dx.doi.org/10.1016/j.bpg.2021.101756]

[3] Francis, P.; Navarro, VJ. Drug-Induced Hepatotoxicity. [Updated 2024 Sep 10]. In: StatPearls [Internet]. Treasure Island (FL): StatPearls Publishing. *N. Engl. J. Med.,* **2025**. Available from: https://www.ncbi.nlm.nih.gov/books/NBK557535/

[4] Pfeifer, P.; Mathys, B.; Alan, E.M.; Bracht, T.; Choinowski, S.; Adorjan, K.; Klöppel, S. Therapeutic drug monitoring of psychotropic drugs in inpatient psychiatry: Local practice and improvement. *Psychiatry Clin Psychopharmacol,* **2025**, *35*(4), 389-395. [http://dx.doi.org/10.5152/pcp.2025.24981]

CHAPTER 21

Vitamins and Trace Elements

Abstract: Vitamins and trace elements play crucial roles in maintaining human health by actively participating in various metabolic processes. These micronutrients are indispensable for enzyme function, antioxidant defence, and immune system support. Deficiencies or excesses of these micronutrients can lead to a wide range of health issues, including anemia, immune dysfunction, and neurological disorders. This article thoroughly explores the practical applications of vitamins and trace elements, emphasizing their biochemical roles, clinical implications, and the interpretation of relevant laboratory tests. It provides a comprehensive overview of the diagnostic challenges and management strategies associated with micronutrient imbalances.

LEARNING OBJECTIVES

After reading this chapter, one should be able to:

- Describe the importance of trace elements and vitamins to metabolic processes and well-being, the mechanisms of their actions, and the consequences of deficiency and overload states.
- Describe the causes, clinical signs, and symptoms of the following aspects of disorders of trace element and vitamin metabolism.
- Describe which laboratory investigations are important in their detection, diagnosis and management.
- Describe the following specific laboratory investigations important to the study of trace elements and vitamins.
- Recognize analytical methods available for their measurement.
- Select laboratory investigations and interpret the analytical results in the context of clinical signs and symptoms.

Simmi Kharb

KEY CONCEPTS

Key Concepts and Applied Aspects of Trace Elements and Vitamins

Trace Elements and Vitamins: Mechanisms of Actions and the Consequences of Deficiency and Overload States (Table 1) [1, 2]

Table 1. Trace elements and vitamins: applied aspects.

Category	Key Concepts	Applied Aspects
Trace Elements	**Importance and Functions**	**Clinical Relevance and Applications**
	Iron (Fe)	Essential for hemoglobin, myoglobin, and enzyme function
	Zinc (Zn)	Important for immune function, enzyme activation, and DNA synthesis
	Copper (Cu)	Cofactor for enzymes involved in energy production and neurotransmitter synthesis
	Selenium (Se)	Component of antioxidant enzymes (glutathione peroxidase), thyroid hormone metabolism
	Iodine (I)	Crucial for thyroid hormone synthesis and regulation of metabolism
	Manganese (Mn)	An enzyme cofactor in the metabolism of amino acids, cholesterol, glucose, and carbohydrates
	Chromium (Cr)	Enhances insulin action, involved in carbohydrate, fat, and protein metabolism
	Molybdenum (Mo)	Cofactor for enzymes in sulfur amino acid, purine, and pyrimidine metabolism
	Mechanisms of Action	
	Cofactor Functions	Many trace elements act as cofactors for enzymes, enhancing catalytic activity
	Antioxidant Roles	Selenium and zinc are critical for antioxidant defense systems
	Hormone Synthesis	Iodine is essential for thyroid hormone production, and iron is important for erythropoiesis
	Immune Function	Zinc, selenium, and copper are vital for maintaining a healthy immune system

(Table 1) cont.....

Category	Key Concepts	Applied Aspects
Vitamins	**Importance and Functions**	**Clinical Relevance and Applications**
	Vitamin A (Retinol)	Essential for vision, immune function, and cell growth
	Vitamin B1 (Thiamine)	Coenzyme in carbohydrate metabolism and nerve function
	Vitamin B2 (Riboflavin)	Part of the coenzymes involved in energy production and cellular function
	Vitamin B3 (Niacin)	Involved in DNA repair, energy production, and cellular metabolism
	Vitamin B5 (Pantothenic Acid)	Coenzyme in fatty acid metabolism and synthesis of coenzyme A
	Vitamin B6 (Pyridoxine)	Cofactor in amino acid metabolism, neurotransmitter synthesis, and hemoglobin production
	Vitamin B7 (Biotin)	Coenzyme in the metabolism of fatty acids, amino acids, and glucose
	Vitamin B9 (Folate)	Critical for DNA synthesis and repair, important in pregnancy for fetal development
	Vitamin B12 (Cobalamin)	Essential for red blood cell formation, neurological function, and DNA synthesis
	Vitamin C (Ascorbic Acid)	Antioxidant, important for collagen synthesis, immune function, and absorption of non-heme iron
	Vitamin D (Calciferol)	Regulates calcium and phosphate metabolism, crucial for bone health
	Vitamin E (Tocopherol)	Antioxidant and protects cell membranes from oxidative damage
	Vitamin K	Essential for blood clotting and bone metabolism
	Mechanisms of Action	
	Coenzyme Roles	Many vitamins function as coenzymes in metabolic reactions (*e.g.*, B vitamins in energy metabolism)
	Antioxidant Activity	Vitamins C and E protect cells from oxidative stress
	Hormone-Like Functions	Vitamin D acts like a hormone in calcium and phosphorus homeostasis
	Gene Expression Regulation	Vitamins A and D regulate gene expression
	Immune Function	Vitamins A, C, D, and E play roles in maintaining and regulating the immune system

(Table 1) cont.....

Category	Key Concepts	Applied Aspects
Clinical Relevance and Applications	**Deficiency Disorders**	
	Iron Deficiency Anemia	Characterized by fatigue, weakness, and pallor, treated with iron supplementation
	Zinc Deficiency	Leads to impaired immune function, hair loss, and growth retardation
	Iodine Deficiency	Causes goiter and hypothyroidism, particularly in pregnant women, leading to cretinism in offspring
	Vitamin A Deficiency	Results in night blindness and increased infection risk
	Thiamine Deficiency (Beriberi)	Causes neurological and cardiovascular symptoms, common in alcoholism
	Riboflavin Deficiency	Leads to ariboflavinosis, with symptoms like sore throat, redness, and swelling of the mouth
	Niacin Deficiency (Pellagra)	Characterized by dermatitis, diarrhea, and dementia
	Vitamin B12 Deficiency	Results in megaloblastic anemia and neurological issues
	Vitamin C Deficiency (Scurvy)	Causes bleeding gums, joint pain, and impaired wound healing
	Vitamin D Deficiency	Leads to rickets in children and osteomalacia in adults
	Vitamin K Deficiency	Causes bleeding diathesis due to impaired clotting
	Excess and Toxicity	
	Iron Overload (Hemochromatosis)	Can cause organ damage, particularly in the liver, heart, and pancreas
	Vitamin A Toxicity	Leads to liver damage, headache, and increased intracranial pressure
	Vitamin D Toxicity	Causes hypercalcemia, leading to vascular and tissue calcification
Laboratory Testing and Monitoring	**Analytical Methods**	
	Spectrophotometry	Measurement of trace elements and vitamins
	Immunoassays	Detection of specific vitamins (*e.g.*, Vitamin D, Vitamin B12)
	Atomic Absorption Spectroscopy (AAS)	Used for measuring concentrations of trace elements
	High-Performance Liquid Chromatography (HPLC)	Separation and quantification of vitamins
	Interpretation of Results	
	Reference Ranges	Establishing normal levels for different populations
	Deficiency and Toxicity Thresholds	Identifying levels indicating deficiency or toxicity
	Clinical Correlation	Integrating laboratory results with clinical symptoms and dietary intake

DISEASE

Biochemical Basis, Relevant Lab Tests, Interpretation of Results, and Applied Aspects

Laboratory Investigations for Trace Elements and Vitamins, and Interpretation of the Analytical Results (Table 2) [1 - 3]

Table 2. Biochemical basis, relevant lab tests, interpretation, and applied aspects of trace elements and vitamins.

Category	Biochemical Basis	Relevant Lab Tests	Interpretation of Results	Applied Aspects
Essential Metals	Necessary for biological functions (*e.g.*, Fe, Zn, Cu)	Serum levels, RBC levels, and urine tests	Reference ranges, deficiency, and toxicity levels	Deficiency can lead to anemia (Fe), immune dysfunction (Zn), Wilson's disease (Cu). Excess can be toxic.
Non-Essential Metals	Metals not required for biological functions (*e.g.*, Pb, Hg)	Blood lead levels and urine mercury levels	Toxic levels indicating exposure	Lead and mercury poisoning can cause neurological damage.
Fat-Soluble Vitamins	Stored in body fat and essential for various functions	Serum levels and fat-soluble vitamin panel	Deficiency or toxicity levels	Over-supplementation can lead to toxicity (*e.g.*, hypervitaminosis A).
Vitamin A and Carotene	Important for vision and immune function	Serum retinol and carotene levels	Deficiency (night blindness) and toxicity (hypervitaminosis A)	Excess intake can cause liver damage and teratogenic effects.
Vitamin D	Regulates calcium and phosphorus metabolism	25-hydroxyvitamin D levels	Deficiency (rickets, osteomalacia), and toxicity (hypercalcemia)	Essential for bone health, immune function, and cancer prevention.
Vitamin E	Antioxidant-protecting cell membranes	Serum alpha-tocopherol levels	Deficiency (neuromuscular problems) and toxicity (bleeding)	Important for skin health, immune function, and preventing oxidative damage.
Vitamin K	Important for blood clotting and bone metabolism	Prothrombin Time (PT), INR, and serum phylloquinone levels	Deficiency (bleeding disorders) and toxicity (hemolytic anemia in newborns)	Essential for coagulation, preventing hemorrhagic disease in newborns.

(Table 2) cont.....

Category	Biochemical Basis	Relevant Lab Tests	Interpretation of Results	Applied Aspects
Folate Metabolism and Function	DNA synthesis, repair, and methylation	Serum folate, RBC folate, and homocysteine levels	Deficiency (megaloblastic anemia, neural tube defects)	Critical during pregnancy for fetal development and preventing neural tube defects.
Genetic Disorders of Copper Metabolism	Conditions affecting copper absorption and transport	Serum copper, ceruloplasmin levels, and genetic testing	Wilson's disease (excess copper) and Menkes disease (copper deficiency)	Managing copper levels through chelation therapy or copper supplements.
Genetic Disorders of Iron Metabolism	Disorders affecting iron absorption, storage, and utilization	Serum ferritin, transferrin saturation, and genetic testing	Hemochromatosis (iron overload) and anemia (iron deficiency)	Regular phlebotomy for hemochromatosis, iron supplementation for deficiency.
Potential Vitamin Toxicity	Over-supplementation leading to harmful effects	Serum vitamin levels and clinical evaluation	Symptoms of hypervitaminosis (A, D, E, K)	Monitoring supplement intake to avoid toxicity, particularly fat-soluble vitamins.
Toxic and Non-Toxic Metals	Metals with toxic effects *versus* those with no known toxicity	Blood and urine metal levels and hair analysis	Elevated levels indicating exposure	Chelation therapy for toxic metal exposure, ensuring environmental safety to prevent exposure.
Vitamin B12 Absorption, Metabolism, and Function	Coenzyme in DNA synthesis, nerve function	Serum B12, methylmalonic acid, and homocysteine levels	Deficiency (megaloblastic anemia and neurological symptoms)	Essential for neurological function and hematopoiesis, supplementation for pernicious anemia or dietary deficiency.
Water-Soluble Vitamins	Not stored in the body and need regular intake	Serum levels and urine excretion	Deficiency and excess levels	B vitamins (energy metabolism), Vitamin C (antioxidant, collagen synthesis). Frequent intake is needed.
B Group Vitamins	Energy production and nervous system function	Serum B1, B2, B6, B9, and B12 levels	Deficiency (various symptoms depending on specific vitamin)	B1 (beri-beri), B3 (pellagra), B9 (folate deficiency), B12 (pernicious anemia). Regular intake essential.

(Table 2) cont.....

Category	Biochemical Basis	Relevant Lab Tests	Interpretation of Results	Applied Aspects
Vitamin C	Antioxidant and collagen synthesis	Serum ascorbic acid levels	Deficiency (scurvy) and toxicity (gastrointestinal distress)	Important for immune function, antioxidant protection, and collagen synthesis. Prevents scurvy.

DISORDERS

Interpretation of Analytical Results and Analytical Methods Available for the Laboratory Tests

Laboratory Investigations for Trace Elements and Vitamins, Analytical Methods Available and Interpretation of the Analytical Results (Table 3) [1 - 3]

Table 3. Laboratory investigations for trace elements and vitamins, analytical methods available, and applied aspects.

Analyte	Relevant Laboratory Tests	Analytics Methods	Applied Aspects
Arsenic	Blood and urine arsenic levels	Atomic absorption spectrometry and ICP-MS	Detect exposure and monitor levels in occupational settings or in suspected poisoning cases.
Cadmium	Blood and urine cadmium levels	Atomic absorption spectrometry and ICP-MS	Assess exposure, monitor workers in industries that use cadmium, and manage toxicity.
Ceruloplasmin	Serum ceruloplasmin levels	Immunoturbidimetric assay and nephelometry	Screening for Wilson's disease and monitoring copper metabolism disorders.
Delta Aminolevulinic Acid (ALA)	Urine ALA levels	Spectrophotometry and chromatography	Marker for lead exposure or porphyrias, monitor occupational exposure.
Ferritin	Serum ferritin levels	Immunoassay and ELISA	Assess iron stores, diagnose iron-deficiency anemia, and monitor iron therapy.
Folate	Serum or RBC folate levels	Microbiological assay and HPLC	Evaluate folate deficiency, monitor supplementation during pregnancy or malabsorption disorders.
Homocysteine	Plasma or serum homocysteine levels	HPLC and immunoassay	A marker for cardiovascular disease risk, monitor treatment efficacy in hyperhomocysteinemia.

Analyte	Relevant Laboratory Tests	Analytics Methods	Applied Aspects
Intrinsic Factor Antibody	Serum intrinsic factor antibody	ELISA and immunoblotting	Aid in diagnosing pernicious anemia and assessing intrinsic factor deficiency.
Specific B Vitamin Status	Serum levels of B1 (thiamine), B2 (riboflavin), B6 (pyridoxine), and B12 (cobalamin)	HPLC and microbiological assays	Assess deficiency, guide supplementation strategies, and monitor treatment responses.
Lead	Blood and urine lead levels	Atomic absorption spectrometry and ICP-MS	Screen for lead poisoning, monitor occupational exposure, and assess environmental exposure risks.
Mercury	Blood and urine mercury levels	Atomic absorption spectrometry and ICP-MS	Detect exposure and monitor levels in occupational settings or in suspected poisoning cases.
Methylmalonic Acid (MMA)	Serum or urine MMA levels	HPLC and GC-MS	A marker for vitamin B12 deficiency, monitor treatment response in pernicious anemia.
Serum Iron	Serum iron levels	Colorimetric assay and spectrophotometry	Evaluate iron status, diagnose iron deficiency or overload disorders.
Urinary Iron	Urinary iron levels	Colorimetric assay and spectrophotometry	Assess iron excretion, aid in diagnosing iron metabolism disorders.
Serum Copper	Serum copper levels	Colorimetric assay and atomic absorption spectrometry	Assess copper metabolism disorders and monitor Wilson's disease.
Transferrin	Serum transferrin levels	Immunoturbidimetric assay and nephelometry	Evaluate iron binding capacity and diagnose iron deficiency anemia.
Vitamin B12	Serum vitamin B12 levels	Microbiological assay and chemiluminescent assay	Assess B12 deficiency, monitor treatment responses, and evaluate pernicious anemia.
Vitamin D	Serum 25-hydroxyvitamin D levels	Immunoassay and LC-MS/MS	Screen for deficiency, monitor supplementation, assess bone health, and calcium metabolism.

SUMMARY

Vitamins and trace elements are crucial for various body functions. Deficiencies can cause health issues, while excessive intake can lead to toxicity. Lab tests are essential for diagnosis and management. Advanced techniques like mass spectrometry have improved accuracy.

EXERCISES

Describe briefly:

- Functions and consequences of vitamin D deficiency.
- Role of iron and consequences of its deficiency.
- Absorption and symptoms of vitamin B12 deficiency.
- Clinical significance of serum zinc levels and associated conditions.
- Laboratory methods for assessing nutrient levels and their limitations.
- Implications, diagnosis, and management of hypervitaminosis A.
- Relationship between selenium levels and thyroid function.
- Biochemical markers for assessing copper deficiency and toxicity, along with clinical manifestations.
- Effects of vitamin K deficiency on blood coagulation and common causes.
- Role of folate in DNA synthesis and effects of its deficiency during pregnancy.

CONCLUSION

Understanding the functions, deficiencies, and toxicities of vitamins and trace elements is crucial for healthcare professionals. Effective management involves accurate assessment, supplementation, and dietary interventions. This study will equip readers with the skills to diagnose, manage, and prevent related disorders, ultimately improving patient care.

REFERENCES

[1] National Research Council (US) Committee on Diet and Health. Diet and Health: Implications for Reducing Chronic Disease Risk. Washington (DC): National Academies Press (US); **1989**. 14, Trace Elements. Available from: https://www.ncbi.nlm.nih.gov/books/NBK218751/

[2] Gerasimidis, K.; Bronsky, J.; Catchpole, A.; Embleton, N.; Fewtrell, M.; Hojsak, I.; Indrio, F.; Hulst, J.; Köglmeier, J.; de Koning, B.; Lapillonne, A.; Molgaard, C.; Moltu, S.J.; Norsa, L.; Verduci, E.; Domellöf, M. Assessment and Interpretation of Vitamin and Trace Element Status in Sick Children. *J. Pediatr. Gastroenterol. Nutr.*, **2020**, *70*(6), 873-881. [http://dx.doi.org/10.1097/MPG.0000000000002688] [PMID: 32443051]

[3] Barr, A.J. The biochemical basis of disease. *Essays Biochem.*, **2018**, *62*(5), 619-642. [http://dx.doi.org/10.1042/EBC20170054] [PMID: 30509933]

SUBJECT INDEX

A

B

C

D

E

F

G

H

I

L

M

N

O

P

T

U

V

www.ingramcontent.com/pod-product-compliance
Lightning Source LLC
LaVergne TN
LVHW070946210926
848482LV00002B/114